MY LOVE AFFAIR WITH FEAR

HOW FEAR ENABLED ME TO BECOME WHO I WAS CREATED TO BE

BRAD KILB

◆ FriesenPress

One Printers Way
Altona, MB R0G 0B0
Canada

www.friesenpress.com

Copyright © 2022 by Brad Kilb
First Edition — 2022

Illustrator: Jody Skinner (painting of the cowboy)
Author photo: Roth and Ramberg Photography

All rights reserved.

No part of this publication may be reproduced in any form, or by any means, electronic or mechanical, including photocopying, recording, or any information browsing, storage, or retrieval system, without permission in writing from FriesenPress.

ISBN
978-1-03-913057-9 (Hardcover)
978-1-03-913056-2 (Paperback)
978-1-03-913058-6 (eBook)

1. BIOGRAPHY & AUTOBIOGRAPHY, EDUCATORS

Distributed to the trade by The Ingram Book Company

Table of Contents

Dedication   vii

Foreword   xv

What They Say About Brad   xviii

Preface   xxi

Introduction   1
- My Life's Transformational Event   3
- Mom and Dad   4
- My Family   6
- My Mentorship   8
- My Spiritual Life   9
- My Challenge   9

1. Fear   11
- My Job, My Responsibility   13
- Portaging   13
- My Daughter, or Me   14
- Standing Up for Our Values   16
- REFLECTIONS. Dancing with Fear   19

2. Explore and Discover   25
- The Foundation of Who I Am Today   25
- My Traumatic First Day   26
- My Dairy Bull Teacher   28
- REFLECTIONS. Explore and Discover   29

3. Learning to Welcome Fear   33
- Ridin' the Rails   33
- The Shortcut   37
- Busted   38
- The Clock Shows Four Seconds   42
- REFLECTIONS. Risk Management or Risk Avoidance   43

4. Leaning Into the Unknown — 47
- African Adventures — 47
- Speed Bump — 48
- No Headlight – No Problem — 48
- Juju Priest Caring for His Ancestors — 49
- Juju Priest's Power Over Crocodiles — 51
- Snake Oil Charlatan — 52
- About to be Sacrificed? — 55
- First Night in Camp — 58
- Outward Bound Journal Entries — 59
- Capstone Activity — 61
- REFLECTIONS. Leaning into the Unknown — 62

5. Getting Comfortable with The Uncomfortable — 67
- Mount Baker Releases Her Fury — 67
- Caving Terror — 69
- A Day at Our Retail Outlet — 75
- Schoolyard Fight — 77
- World Indoor Beach Volleyball Championships — 79
- REFLECTIONS. Getting Comfortable with the Uncomfortable — 85

6. Creating the Dream — 89
- Surfing to Save My Life — 89
- Dog Pile — 96
- Swimming with a Humpback — 98
- My Encounter with Sharks — 102
- Croc-Infested Waters — 105
- Controlling Thoroughbred Stallions — 108
- River Rescue — 111
- Speed Kills, but it Sure is Exciting — 118
- REFLECTIONS. Creating the Dream — 121

7. Terrifying Health Discoveries — 125
- Battling Malaria — 125
- Following Dad's Advice — 126
- Flatlining in the Operating Room — 128

The Silent Killer	131
Skateboarding	135
Three Months of Hell	137
REFLECTIONS. Divine Intervention	141

8. Our Creator Will Decide When It's Time — 145

Dealing with Death	145
Lorna – My Mom	147
Albert – My Dad	149
John Ross MacRae – My Nephew	151
Brett – My Son	152
Brad Jr – My Son	156
REFLECTIONS. Our Creator Will Decide When It's Time	158

9. Coaching Decisions — 163

Escaping the Sicilian Mafia	163
Undercover Agents	164
Double or Nothing	167
Casino Float	169
Motocross Nationals	171
Sometimes There's a Cost to Doing the Right Thing	175
World Championships, Rio de Janeiro	177
REFLECTIONS. There's More to Sport Than the Score	178

10. Unpredictable Animals — 189

Italian Cinghiale	189
Runaway Team	192
Wedding Carriage Ride	194
Release the Beast	196
Sometimes Teamwork is Essential	199
Killer Whales	200
REFLECTIONS. Magical Moments with Animals	201

11. Nature's Power — 205

Kayaking the Sea of Cortez	205
Pacific Storm	208
Caught in a Hurricane	213

The Ram River	215
REFLECTIONS. Nature's Power	220

12. Quick Thinking — **225**
Runaway Trailer	225
Caught with My Pants Down	226
What's that Guy Doing in His Pyjamas?	227
Stepping into the Modelling World	228
REFLECTIONS. Quick Thinking,	230

13. My Transformation Strategies — **233**
Forced Actions	236
Accepted Actions	236
Unplanned Actions	242

14. Don't Let Fear Limit Our Potential — **247**

Ever So Grateful — *249*

About the Author — *251*

Dedication

To my wife Bonnie, and my six cherished children:

Jamey, Bryn, Jodi, Brad Jr, Brett, and Justin,

who are gifts from heaven helping me to embrace life fully.

Bonnie

Jamey

Bryn

Jodi

Brad Jr

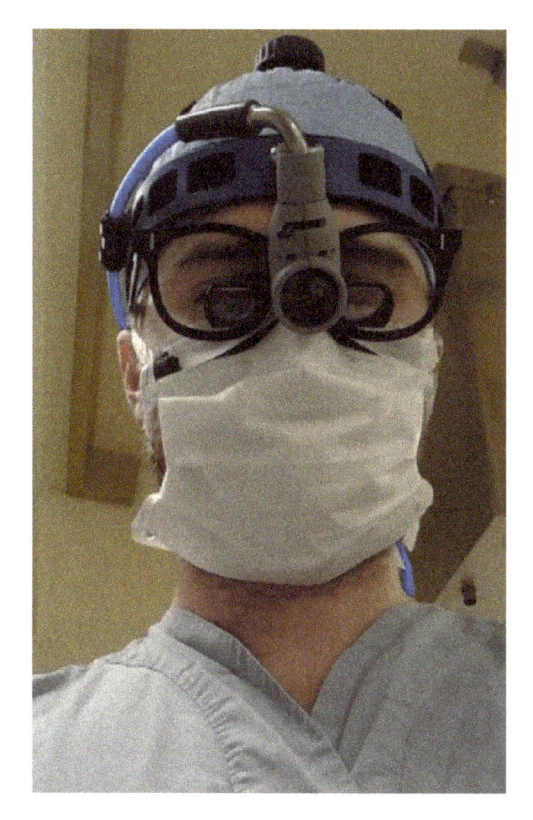

Brett

Justin

Foreword

As a long-time school principal, I've watched great teachers/coaches change "loser" lives into lifelong "winner" lives. They promote confidence, inspire commitment, nurture positive character, and build personal values. They help young people maximise their God-given gifts. In turn, those young people contribute to successful schools and positive communities. Many of them became teachers/coaches.

Twenty years ago, I met the best teacher/coach I've ever known while on a Mexican vacation. Brad and I sat under a beach palapa and began a wonderful friendship. I listened to many stories that made me laugh. Some of his ideas made me a thoughtful educator. A few stories made me cry. Every year since, Brad has been part of my holiday in Mexico. The stories and insights have never stopped. Finally, he's taken the time to put them in print.

I'm going to read Brad's book for two reasons. First, I love to be entertained. Brad is Canada's answer to *Tom Sawyer* – Mom and Dad raised him to lead a full and adventurist life. To Brad, that meant ridin' the rails; bustin' a wild bronc; climbing the highest mountain; facing off with scary predators. Brad personalized "Extreme Sports" as a young man. He continues to live them in his 70s. And, just our luck, every story seems to include humour that will make you laugh.

I wonder how many times our palapa discussions have been interrupted by beach friends, "Tell us about swimming with sharks! What about the runaway trailer, the Juju priest and the crocodiles, the liquor store and the hand grenade?" Those of us on our Mexican beach have turned Brad into the Pied Piper of great stories.

Several of Brad's adventures happened while teaching outdoor survival classes. My favorite involved an avalanche. Brad was swept down a mountain on the tongue of an avalanche. It ended when his well-trained students dug him out. One student was particularly impressed, "WOW! Thanks for that great demonstration, Brad."

I hope my grandchildren give special attention to my second reason for reading Brad's book. Brad shares his insights and strategies for leading "a good Christian life" throughout the book—they speak to athletes, teachers, coaches, parents, and grandparents. I suspect each of my grandchildren will find unique ideas that most relate to their personalities and lives. I'm confident they will lead better lives and be better people for reading this book.

Although this never came up in our discussions, I'm aware that Professor Brad's been inducted into the University of Calgary *Teaching Excellence Hall of Fame.* He's a best-selling author and an internationally acclaimed film producer. I'm very cognizant that Brad is held in such high regard. He's also found time to be a loving son, husband, father, and grandfather.

Brad's greatest achievement was when he convinced a former student and athlete to marry him. Bonnie's loved and supported this guy through many of his adventures. I suspect he's with us today because there were times when she said, "No!"

Tal Guppy, Former Principal, Kent School District, Washington

Brad is a man with many talents, gifts, interests and accomplishments. I am the beneficiary of what I would consider his most precious gift … his ability to see potential and the possibility of greatness in others. The trajectory of my life changed when Brad saw me as a young 17-year-old girl from Saskatchewan with a dream to play volleyball at the highest level. He saw me and chose me to play on the Canadian Junior National Team, and that made all the difference in the world for me. I loved him as my coach and cherish him as my friend today.

Gisele Kreuger – Performance Coach, *The Winner Will*

You lead with a style and ability to pull the best out of people in the most supportive way of anyone I've ever met. It's truly astonishing the effect you have on people, myself included.

Erin Mcintosh, University Coach

I don't think I can adequately explain how much Brad's mentorship, friendship, and presence in my life has meant to me. Brad has changed the course of my life and I thank God deeply for him. Brad shows a genuine care for those around him, modelling vulnerability and true friendship. He leads by example with an understanding and emotional/social intelligence, always aware of how I'm feeling and what I need. I think perhaps the greatest impact he's had on me comes from simply being around him—catching his positive intensity and spirit of joy.

Brad lives a life of gratitude and lets that flow out of him freely to influence those around him with confidence and wisdom. His example of being brave, taking on new challenges, servant leadership, and living each day to the fullest inspires each of us by honouring God as he uses his gifts to the max. Brad, with the most knowledge, wisdom, experience, and resiliency of anyone on our team, still models a growth mindset every day. Perhaps it's one of the biggest intangible and unseen gifts he brings to our team.

Brad's been a prophetic influence in my life—someone who sees gifts and potential in me and is "calling me higher" into its reality. He has an amazing ability to look into my life, discovering and articulating gifts and areas needing improvement. His mentorship has earned such trust to speak, instruct, and model. I am so grateful for him as a role model in my life.

Colin Kubinec, University Director of Athletics, Former University Coach

What They Say About Brad

Colleagues

Brad's experience in the field of canoe instruction and white-water rescue is without peer in the world today. I have been tremendously impressed with the worldwide reputation by which this amazing professional is known.

John Gallagher, Vice-president, *USA National Association for Search and Rescue*

Your words have always been filled with more meaning than you know. I think the Spirit knows exactly what words we need to hear at certain times, and you are so faithful in articulating them to us. These encouraging words help me step into those gifts you see in me!

Mackenzie Schmidt, *University Coach*

University Students

It truly is a treasure having a professor who treats us as a gift to the world rather than just another number. You inspire us to be more and to have limitless expectations for what we can achieve and who we can impact. I am so grateful for the endless life lessons I've learned from you. My whole experience in university has been positively influenced by you more than words can say.

Callie Ann Massee

Your stories, your knowledge, and your wisdom are shaping young hearts and minds in our university. You motivated and challenged me today. You brought new life to my soul and my spirit. You affirmed me in a way that I have never been affirmed before.

Joshua Cairns

Before I took your class, I wasn't a believer that one class could change my life. I now stand corrected, as yours helped me get through the toughest term of my life. Your positivity and drive to become a better person completely blew my mind from day one, and I was soon adapting to that very way of thinking without even realizing it.

Natasha Graver

You truly are a rare educator. I am constantly inspired by the level of personal investment, passion, and individual consideration you bring to each class—your classes are transformative. You have had such a positive impact on my development both personally and professionally.

Kendrick Yee

You've inspired me in ways that no other professor has. You truly care about each of us. I know that our great Creator is so proud and pleased of the wonderful work you have done.

Zach Barfuss

I'm incredibly grateful that you are always looking to improve. I can't begin to tell you how much that means to us students. It creates a psychologically safe environment and allows us to truly lean into our learning. It makes such a large difference to feel that my voice matters and will be heard and valued.

Shae-Lynn Carlson

University Athletes

You truly have and continue to be a MAJOR role model in my life! You've always believed in me, encouraged, and pushed me towards God's best for me. And for that, I am truly grateful.

Somer Bartel

I am continuously amazed at the strength you and your entire family display, even through the death of your two sons - you somehow still make room in your hearts for others even in the toughest of times! The way you walk through these tough times is what I MOST admire you for. You remain soft-hearted, kind, genuine and humble; and you are always building others up!

Brooklyn Dyck

Preface

As my seventy-four-year-old body slides into the cockpit of our double kayak on a self-guided, eight-day circumnavigation of Isla Carmen in the Sea of Cortez, I ask, *Why am I so willing to jump into this adventure, knowing I could come home in a body bag?* With son Bryn, his wife, their ten-year-old son and eight-year-old daughter, we fully understand the devastating dangers we face. The most daunting is our passage across the north end of the island, exposed to high winds, monster waves, rocky shores and unpredictable wildlife.

I believe curiosity is my incentive for adventure. My epic escapades have given me insights into my true self – why I consider fear to be my ally; what my priorities are; who I really am; when I should step into an adventure; where the limits are of my skill sets, mental fortitude and grit; how I can manage personal fear. I can't push aside my fear; it's omnipresent. Managing my fear has enabled me to be more alive, to live a fuller and richer life – as I captain a yacht through a Pacific storm; discover mental toughness as I prepare for the Paris marathon; follow instructions while swimming with a shark in the Red Sea; focus on survival fundamentals as I surf an avalanche in the Rockies; battle a heart attack and five bouts of cancer requiring surgery; fight back tears as I receive the strap in grade seven; feed crocs with a Juju priest in Africa; recover from flatlining in the intensive care unit; swim an underground river while caving in Australia; pull a pilot from a burning plane in Nigeria; wrestle with a grenade-wielding thief; and go standup paddleboarding amongst deadly crocs in Mexico. These frightening experiences shaped who I am today.

Yes, my story is unique, just as yours is unique. Every one of us has a story to tell. Each is different. Each is valuable. I plan to share an intimate story of myself – a person who faces fear every day but chooses to live life to its fullest. In sharing how I've embraced fear, I hope to reveal how I've become the man I am. Hopefully, you'll find my stories engaging and entertaining but also share in the transformative lessons I've learned in my eight decades of life and five decades of research. My desire has nothing to do with wanting you to celebrate me. I hope to inspire you to face your fears and live the fulfilling life your Creator has placed you on this earth to create for yourself.

Each of us goes through a process of evolution and maturity as we age. I'm no different. As I look back, I can see definite changes in my life due to the many experiences I've encountered which have hopefully led me to be more astute in all areas of my life.

I'm a man with a boundless desire for knowledge and a passion for exploring, growing and testing myself. I have a deep-seated hunger for adventure. It's my addiction. My life has taught me this is the best way to grow and become who I was created to be.

Success demands commitment - the willingness to step into a life-transforming risk. For me, it's that desire to grow, to get to know my own boundaries, to assess my capabilities. Sometimes, once we take that first step, there's no turning back. We are swept away into a one-way tunnel –no stepping out, no reverse, no escape, no rescue. We are forced to rely on our own skill sets and grit to come out the other side - walking up to that admired schoolmate asking for a dance; pulling our rappel rope leaving only one way out of our underground cave; facing an enraged school bully waiting to pummel me; entering shark-infested waters as I leave the safety of sandy shores; pushing out of the start gate on an Olympic bobsled run; signing the waiver giving permission to undergo surgery; nodding my head to open the bucking chute while mounted on a snorting bronc.

I become totally and completely captivated in these moments. The world disappears. Nothing else exists. Time slows. All senses focus on the task. Perhaps the most amazing aspect of my adventures is the rare, surreal feeling felt when I've become immersed in the experience as if I belong, rather than participating as a spectator. These are mind-blowing encounters, so hard to replicate. I've been blessed with these moments.

I, like us all, am imperfect and flawed. I'd be the first to admit that I'm not proud of all stages of my life. My life has been a struggle – my body ravaged by disease, my family broken by divorce and my resilience tested by the death of two young sons. However, as tough as these moments have been, my prime emotion is gratitude. I've had such good fortune with moments of joy, love, peace and fulfilment. I believe that God doesn't give us more than we can handle. It's true, as Brandy Halliday says, that 'Imperfect people can experience perfect moments.' Our personal challenges shape stronger individuals.

Some of my adventures are risky, while others are quite mundane. Some are forced, some planned. A few veered off the tracks. Others have evolved because I've been in the right place at the right time. Fear has often been self-inflicted, sometimes uninvited. I see it as an invitation to enter the greatest adventure I've ever known – my life as it's meant to be lived – the way God wants me to live. I've been willing to be vulnerable and face failure, surviving countless close calls with death with unpredictable animals and in the mountains, oceans, underground caves and hospitals. In some cases, the wrong decision could lead to injury or fatality! It's the price I'm willing to pay for that sense of being so acutely in the present, so acutely alive. I've never taken on a new adventure believing I was going to die. Foolhardy risk is not part of my make-up. After accomplishing something very daunting, the elation I feel is an emotion I cherish.

I'm not attracted to adventure because I'm an adrenalin junkie or thrill seeker. It's not an attraction to death, but the opposite – a passion for that sense of insightful aliveness. I want to experience new adventures. I want to step out of the mundane. I want to live in a different way than the norm. I want to step into the unknown. I do not want to live the same life year after year and call it a meaningful lifespan. I continue to grow as I learn to dance with fear, explore, dare to attempt and learn from my mistakes.

Epic adventures demand one's total attention – a total focus and full sensory alert when danger is near. There's no past, no future, no outlying fantasies for what might be or haunting remorse for what could have been. There is only the present – this moment which is utterly encompassing as I focus on my immediate surroundings – the actions I must execute. The victorious moment is fleeting. It's the process that's transforming.

At the same time, I want to utilise my Creator's gifts to the fullest. I want to find a place where I can facilitate growth as I mentor others to discover their purpose – that they are unique with special God-given gifts. My chosen occupations allow me to do just that every day. I'm so grateful God has endowed me with the gifts that allow me to coach, mentor and parent. In these roles, I'm able to bring out the best in those I lead and help them work *through* their fears.

My occupational goals changed numerous times early in my life. In elementary school, the only occupation I considered was that of an NHL hockey player. As I entered the University of Ghana, I studied science to become a dentist. Upon the completion of my Outward Bound course in Nigeria, I determined to return to Canada to start the first Outward Bound School in our nation. At one point in my life, I considered becoming a professional rodeo rider. However, after my fourth child, I became aware that placing myself in danger was not fair to my family, so I hung up my spurs. Then, five decades ago, I slid into the teaching profession. As I look back at my years of mentoring, I'm convinced this is where my Creator wants me to serve. I absolutely love my job.

At our university, I've been inducted into the Teaching Excellence Hall of Fame. My coaching has resulted in numerous national championships in volleyball, white-water paddling and motocross. Coaching Canada's National Volleyball Team and professionally in Japan, Italy and Switzerland has given me renown. As the coach of our university team, I was selected as Canadian Coach of the Year by my colleagues. As I look at my life and my accomplishments, I'm proud of my achievements not because of medals, trophies and titles, but because of the small role I've played in facilitating others to 'be all they can be'.

Regardless of my position in life, I want to leave a legacy that speaks of integrity (an open and transparent life aligning with my core values), caring (behaviour reflecting empathy and compassion), facilitating (enabling those I lead to take ownership of their own growth) and mentoring (assisting those I lead to realise their full potential) as I strive to be the best husband, father and mentor possible. National championship rings and hall of fame inductions hopefully give recognition of these traits I value so much. I wish to share how I think and how I process my actions.

Yes, I've failed and been disappointed, but my response has led to moments of self-discovery leading to transformational growth. With any luck, I will inspire you to venture where you have never been before.

Over the past couple of years, I've started to record my adventures. I fully believe my Creator has placed me on earth with unique gifts, and I accept the challenge that He is keeping me alive until I've finished His work. I've quickly discovered that putting stories to paper is much more difficult than anticipated. However, I'm loving the challenge.

During the writing process, I've become increasingly aware of the multitude of my adventures. Two people have been a part of most of them. My first wife, Margie, and my present wife, Bonnie, have had to endure my adventurous spirit. They do not need any reminders! I do need them to know how much I appreciate their never-ending support. Without their partnership, I would have been unable to enjoy the lifestyle I've lived.

Along with my faith, the most important component in my life is my willingness to be curious; to probe; to step out of my comfort zone; to be vulnerable, and to not be afraid to fail. The greatest moments in my life have been the most difficult ones – those times I've learned and grown. I need to feel uncomfortable in order to grow. Life means nothing if I choose to relax in my comfort zone.

My faith inexplicitly brings me comfort as I struggle with fear – as I grieve the loss of my sons; love my addicted children with unconditional love, forge our blended family into the best it can be, make difficult decisions, plunge into dangerous adventures and share impactful words with broken friends. Do I believe I receive answers to prayers? Absolutely, but often not in my timeframe!

Introduction

'Be fearless in the pursuit of what sets your soul on fire.'   JENNIFER LEE

Author Brad

Photo: Impact Magazine

All of us have fearful challenges to face. My challenges have come from many walks of life – relationships with my blended family; coaching at local, national, international and professional levels; coping with numerous health issues requiring surgery; helping my children through drug and alcohol addictions and permanent brain damage; and coping with the deaths of two of my sons, aged twenty-eight and forty-six. These challenges came at the most unexpected times and have left reminders of how transformative and, at the same time, how blessed my life has been – two incredible wives; six amazing children; delightful, energetic grandchildren; a loving, blended family; a teaching and coaching career; a survivor of five bouts of cancer, a heart attack and flatlining in the intensive care unit; and a global explorer travelling to sixty-seven different countries, residing in seven.

My Life's Transformational Event

I was preparing for a career in dentistry as I entered my first year of science at the University of Ghana, West Africa. My ambition was to spend my life peering into the oral orifices of others. My dream of dentistry included setting my own working hours and making wads of money. I would have spare time to embark on those outdoor adventures I so love.

While in Accra, I came across a human-interest story about the Nigerian Outward Bound School. This British International Organisation claimed to change lives through challenge and discovery – to develop character and leadership. Wow! How could I miss this opportunity for personal growth while participating in outdoor challenges? I immediately shot off a letter requesting information on how I could enrol in this twenty-eight-day course.

I began my life-changing journey hitch-hiking thousands of kilometres across four African countries to base camp in Jos, Nigeria. It was 1961 and as a penniless white student, hitch-hiking was not only cheap but safe.

The course far surpassed any expectations I had harboured. This was my crossroads – my life's transformational moment. My decision would direct the remainder of my life. Should I pursue my dreams of assisting patients in living healthier lives, or should I follow a new path and prepare for a fulfilling career enabling me to follow my passion every day – inspiring youth to challenge themselves, grow and reveal their God-given gifts? After decades of teaching, coaching, guiding, instructing and mentoring, I can say that fork in the road was the best decision of my life.

Mom and Dad

Some of us owe everything we have to the bank accounts our parents passed on to us – second-generation wealth. For me, it has nothing to do with financial wealth. It's the exemplary lives Mom (Lorna) and Dad (Albert) lived. It's an incalculable debt for which I will forever be grateful. They were the most loving and committed couple I've ever seen. They absolutely adored each other.

Mom's love poured into each one of us – Loral (my adopted sister, eight years younger), Brian (my younger brother by two years), me and Dad. It filled my heart with joy as I watched the love between my mother and father. They are such a perfect example of loyalty, support and an unequivocal partnership. A former legal secretary, Mom gave up her profession to be a stay-at-home mother and homemaker. She was always there; hugs as we left our apartment; smiling face and hugs when we returned home; and paying attention to our every need while home. I could not have asked for a more compassionate and caring mother.

A stay-at-home mom, she was never too busy to jump in and participate in my activities or answer my questions: *Mom, should I kiss my girlfriend with my eyes open or closed?* She knew exactly how I should respond and behave.

Her love carried beyond our family. She constantly searched for ways to welcome strangers and friends, no matter the circumstances, particularly if those people looked needy or lonely. If she noticed a visitor walk through our church doors, she would immediately welcome them and invite them over for lunch. Her love knew no limits. It seemed as if Mom believed God had placed her on our planet to let His love flow through her. What a beautiful example she was.

At 1.85 metres tall with handsome looks, Dad walked into the room and demanded attention immediately. His muscular body – that of the professional athlete he was (football and lacrosse) – filled the space with dignity. There was no smoke and mirrors with Dad. He was fearless as he said what he thought and backed up his words with action. He was very clear on his values, never veering from them. His work ethic was amazing – no job too hard or too dirty to tackle. If he could do it himself, he dove in. If not, he was not too proud to seek assistance. Although I pushed back against his strict restraints at the time, I know they made me the disciplined person I am today. He always had time to spend with me – camping, canoeing, coaching, just going into the backyard to play catch.

Dad was the technical director at a large composite high school. He was a champion of the underdog, working hard to elevate the prestige of the trades in a society that placed more esteem on university-bound programmes. He feared no one nor any scenario as he was stalwart in his support of blue-collared folk. As a teenager, he was forced to leave school to earn a salary as a toolmaker to help support his family. He was dissatisfied with his education having been cut short. So twice a week after, teaching all day, he drove 100 kilometres to complete his post-secondary Education Degree at McMaster University.

They are my heroes, my role models, my inspiration. They were never afraid of unchallenged barriers and pushed constraints in every area of their lives, personally and professionally. There is no question they're the ones who instilled my sense of curiosity, fearlessness, exploration, hard work and discovery. They displayed the kind of courageous character I hope to emulate – hard-working with absolute integrity. I revere them with a mixture of awe, admiration and adulation – a life that fills me with wonder, amazement and gratitude – emotions I've never lost.

Parents, Albert and Lorna

My Family

I fully comprehend I've been privileged with innumerable blessings. The most rewarding has been parenting our six children. One of the most encouraging questions we get is, 'How did you raise such incredible children?' It's a question that must be answered modestly. There are so many pieces to the puzzle. We discover new ones every time we talk with our kids. We've always stressed that each child plays the most important role in their own success. I believe the ultimate answer is *to be the examples of the adults we want our children to become.*

In our home, we make it a point to spend time discussing important topics – living a healthy life; making ethical decisions; discovering our God-given gifts; contributing in small ways; creating and articulating knowledge; dealing with death; leaving a legacy; and living a life of faith.

Taking a page from our coaching background, we gather in front of our fireplace on a cold, winter day to explore our family mission statement. We each contribute. After each opinion is written on a large whiteboard, we begin a discussion. We encourage our kids to stand up for their beliefs. It's expected they will learn negotiation skills as, with open minds, they actively discuss the thoughts brought forward. A week or two later, we set up the whiteboard again to further refine our core values and mission. We continue until we've agreed upon our document.

Our Family Mission is to
*Create an environment where each of us can find support
and encouragement in achieving our life goals.
Respect and accept each person's unique personality and talents.
Promote a loving, kind and happy atmosphere, where family,
friends and guests find joy, comfort, peace and happiness.
Support family endeavours that better society.
Maintain patience through understanding.
Always resolve conflicts with others rather than harbouring anger.
Exercise wisdom in what we choose to eat, read, see and do.
Appreciate the grace of God and worship together … forever.*

Life includes a series of passageways. We choose our paths. Margie was my partner during my earliest career years as a Christian youth minister, and mother to our four remarkable kids – Jamey, Bryn, Jodi, and Bradley. Unfortunately, I started to place more importance on my career than on my family, building credibility and reputation as an elite coach. I eventually won prestigious awards and landed a head coaching position within Canada's National Volleyball programme. Margie's and my goals and interests drifted apart. I didn't spend enough time being a father to my kids; being the husband my wife deserved.

I remember the day I was schoolyard supervising. I spotted daughter Jamey. Fear struck me with the force of a lightning bolt! I knew Margie and I could not continue our relationship, but *what about the kids? How will this divorce impact the kids I love so much?* I stood terrified. *Do my kids and Margie have to be a package deal? Why can't I end our marriage but keep my kids?* Our relationship ended in divorce.

As I reflect, I believe it's difficult to have a career/family life balance. Those of us, male or female, who strive to be uber parents or corporate superstars must prioritise one above the other. It seems that one of these must be the victim!

I'm now married to Bonnie, a wife who has been a most remarkable partner for me, and an amazing mother to our two boys – Brett and Justin. Bonnie has taught me how to live life to the fullest despite the cruel curves we've had thrown at us. Day in and day out, she encourages me to create and follow my dreams as she strives to be the best version of herself. She's taught me to refuse to give in to the emptiness that fills my heart as we find meaning beyond Brett and Brad Jr's deaths – we cannot lose the lives we've been given despite the loved ones we've lost. I've been blessed with six amazing children – it's the perfect number for a volleyball team – time to say *enough*.

Divorce is never good, but Bonnie, Margie and my kids have created a positive, blended family. It's not been easy for any of us. I believe the root of our family cohesion is the unconditional love of we parents, and the respect the kids have for one another. Our prickly conversations take their toll.

My Mentorship

I want those I lead and you, my readers, to explore your inner selves – discover your gifts; unearth the role your Creator has created you to be; and step in bravely to live that fulfilling life.

What are your core values? Who are you? What brings you joy? What gifts do you bring to the table? What are your strengths and talents? How do you impact others? How are you perceived by others? What career direction have you identified for yourself? Why do you want this direction? Are your daily activities in line with your direction? Are you committed to putting in the work required to become who you wish to be?

My challenge is to inspire you to initiate a direction that will lead to transforming you into the masterpiece God created you to be – a fulfilling life maximising your potential. I can't do that for you. I endeavour to share the tools I've utilised, but I cannot impose my desires upon you. I tried to resolve the horrors facing my addicted children. It didn't work! I could only guide them to their own self-discovery, to accept their addiction and to take brave, personal steps to overcome it. I'm so proud of them for doing so.

Perhaps the biggest task we face as humans are those responsibilities as a partner and parent. Our education system offers little advice. We are thrust into these roles with little or no training. We're forced to stumble our way through both roles. I dare say most of us parent the way we were parented; love the way our parents loved; and discipline the way they disciplined – unless we're able to take the brave steps to break free of a toxic family culture we cannot condone nor perpetuate. Although I had no choice in choosing my parents, I feel immensely blessed by their examples. I strive to emulate how Mom and Dad parented and loved each other.

My faults and imperfections have certainly become evident as I plunged into marriage and my parenthood – I've made more mistakes than the captain of the *Titanic*! My tales are simply narrations of my own setbacks and successes – of how I learned to welcome the fear within the challenges I have so often faced. My life has been so richly blessed as I strive to use failures as my greatest learning opportunities. I thank you for allowing me to share my lessons with you.

My Spiritual Life

My lessons have encompassed all areas of my life. Following my circumnavigation of the globe, I stepped into my role as a youth minister. As a Christian, I felt it was important to 'earn the right' to share my belief. Consequently, I spent many days building relationships with high school students at Pioneer Ranch Camp.

Then, in 1972, I focused on my coaching career, neglecting my spiritual life. On the invitation of coach Kubinec in 2013, I accepted a position as a mentor coach at Ambrose Christian University, and my faith once again grew. How blessed I am that God brought me back into a relationship with Him. It's truly amazing how life sometimes comes full circle – I abandoned my faith for sport, and sport brought me back.

As I reflect on my life's epic events – medical setbacks that could've/should've ended my life, adventures that could have ended in catastrophe and peace of mind following my two sons' deaths – I wonder, why is my life so privileged? Is it coincidence? Is it luck? Is it solely due to my efforts? No. I'm 100% convinced it's my faith – that God has and is playing a role in my life. I'm so grateful.

My Challenge

My challenge to myself, those I mentor, and you my readers, is to be able in the last days of our lives to look back and claim, *I've achieved the purpose for which I've been placed here on earth to accomplish.*

Hopefully, my story will entice you to fearlessly engage in endeavours that will help you grow as individuals and discover more of yourselves.

1. Fear

'I lost my fear and gained my whole life.' CLEO WADE

Just one metre separates us …. a 14-metre whale and myself. With fear raging in every cell in my body, I notice his eye scanning my body. *What is he thinking? What will he do?* It's not the first time I've wrestled with fear, nor will it be the last.

Fear is an emotion every one of us experiences in so many different situations – public speaking, getting a needle, interviewing for a job, facing a physical challenge, striving for sobriety or going to the dentist. It acts as a warning system. Simply insisting, 'Don't be afraid. C'mon, be brave', is not the answer. Grappling with fear is a lifelong struggle. It's not an emotion that disappears, but rather one that stays with us throughout our ordeals. Of course, if our fear is life-threatening, we must back off. But, if it's presenting a learning arena, I've learned we should seize that opportunity to 'dance with fear' – to invite it along on our trials, assisting us in focusing our physical and psychological abilities to conquer the challenge. I don't want fear to rob us of a transformative learning experience.

Fear is natural and good. I look at fear in the same light as pain. Both are warning systems – telling us to pay attention. Both systems can invite us to proceed or advise us to stop. As we touch a hot stove, our pain initiates an immediate reaction. On the other hand, as we work out on a rowing machine, pain is something we must battle through to gain progress. Fear can initiate the same decision – proceed or withdraw? Feeling fear is not my decision, but it is my decision *what I do* with that fear. It can shackle us, holding us back from what life has to offer. When we are fearful, we tend to insulate and protect ourselves from the unknown. As parents, we must create a safe and supportive environment, an approach whereby our kids are willing to take risks knowing we have their back as they utilise their skills to be victorious.

We cannot step into a frightening event and ignore our fear. We must face the challenge while afraid. Learning to dance with fear is exemplified in the bullfighting ring – something I witnessed in Barcelona. As the matador stands in the middle of the ring and the gate of the bull pen is flung open, I imagine he experiences fear as he faces the raging bull. The first act gives the matador the opportunity to dance with the bull as he uses the *capote* (a large cape) to learn of the bull's fighting tendencies and temperament. Does the bull execute long, smooth charges, or does he twist dangerously? Does he show a preference in the use of either horn, or does he attack equally from both sides? Does the bull appear to have poor vision which could signal an inability to follow the path of the cape? These early observations assist the matador in being successful in the final stage of the fight. Instead of hoping the fear will disappear, the matador intentionally dances with fear to learn how to overcome his adversary. Similarly, I've learned to lean into the challenge, focusing on the principles that will dictate success.

Curiosity entices me into fearful undertakings – it's what gets me into trouble at times. But it also opens the door for discovery leading to growth. I'm not sure where I developed my curiosity. I believe some of it is innate (perhaps from my Viking heritage), and some learned. It must be practiced. I have always had an inquisitive mind, but I believe my curiosity's been enhanced as my parents allowed me the space to explore new adventures – pushing my boundaries in unfamiliar and uncomfortable situations. There was no hovering, only encouragement and support.

Transformative events can have an immense impact on our lives. Some of these life-changing events will be willingly entered, while others will be forced upon us or simply occur because we are in the right place at the right time. All these encounters demand courage – utilising fear as a partner as we lean in. How do we muster the courage to step into the unknown *with* fear? In my experience, as soon as fear jumps into my brain, I must assess the situation. Courage is based on an honest evaluation of the situation – my skill set, past experience, confidence and acceptance of who I am (pushing away thoughts of failure, shame, embarrassment). Looking back at our lives, we may be able to identify some of these incidents which may occur at any time. If I can accept failure as part of my journey, I'll appreciate the lessons I learn and find success.

I've experienced numerous transformative experiences. They've formed my values, renewed how I evaluate, altered my priorities and changed my life. Curiosity with an open mind enables new encounters – *if* I have the courage to accept my fears and take that first step.

My Job, My Responsibility

'Brad, are you finished?'

'Yep, the tent's ready to move into, Dad.'

As an eleven-year-old, I'm camping across Canada with my family – fifty-two nights sleeping in a tent with my four family members. Dad delegated a task to each of us. Mom's setting up the kitchen – she's so organised. Brian's blowing up air mattresses – his short attention span amplified with hyperventilating. Loral's preparing sleeping bags – her attention to detail is amazing. Dad's unloading our cartop carrier – he's got a spot for everything (and we'd better put it back). I'm setting up our tent – I love to build. Our pristine campsite's erected in less than twenty minutes.

Dad's set us up for success, using delegation rather than volunteering. Assigning tasks allows Dad to give us tasks he knows we can accomplish. I've been assigned a task, and I work to complete it with perfection. Was I fearful I might not be able to pull my weight? You bet. Those first nights were nerve-racking, but Dad would wander over to assist if I was having trouble. I've learned what it means to be a *team player* – contributing my unique gifts for the benefit of all. I feel so proud to fulfil my role.

Portaging

'Yes, you're finally strong enough to portage, Brad.'

Dad talked about it for many summers on our family canoe trips. He was a powerful, intimidating figure – the epitome of the professional football and lacrosse player he was. Oozing with confidence and earning respect, he towered over me as I looked up past his broad shoulders, his long blonde hair combed straight back. In our first years of tripping, he would portage both canoes while we toted the backpacks.

I stumble across the trail under the 27-kilogram canoe, balancing my load, stepping over roots, looking for places to rest and praying the next waterway will appear under the bow of my canoe. The thwart digs into my neck, pressing against my vertebrae. My tired legs feel like spaghetti as I climb the hill. The mosquitoes bite and suck blood uninterrupted. Sweat rolls down my forehead, the salt stinging my eyes. I dismiss the fear of quitting. *I've got to keep trudging – suck it up. Dad believes I can do it.*

What a feeling of accomplishment as I drop the canoe into the water. In my egotistical mind, I feel it's a rite of passage, *man enough to portage a canoe solo*. I think back at the various feats I consider distinguish me as a 'man's man' during these decades of the 'Marlboro Man', portraying images of a rugged and macho man. *Was it my ego that enabled me to take that first fearful step?*

My Daughter, or Me

'Grandpa, do you know when Mom and Dad will be back?' I ask.

'I have no idea, and to tell the truth, I don't really care.'

I can tell by Grandpa's voice he's in his usual grumpy mood. It's with some worry I wonder what Grandpa will think of our new baby sister as I await the arrival of Mom and Dad. They had applied to adopt a baby girl, and today is the day they'll bring home our infant prize.

Grandpa Kilb immigrated to Canada from Britain early in the twentieth century. Grandpa's father was a Queen's Butcher in Yorkshire, England. With his prestigious position, he became a multimillionaire only to gamble it away at the horse track. To this day, none of us Kilbs have engaged in gambling. My Canadian heritage begins with Grandpa immigrating with six cents in his pocket. He brought his wife Lynda across the pond one year later after landing a job as a toolmaker. With his wife Lynda, they raised my father and his brother Ed. Grandpa's rose garden was the pride of his existence – one of the best in Toronto.

There's no mistaking who is the king of Grandpa's palace. There always seems to be tension when we visit between Dad and Grandpa. As the master of his kitchen, Grandpa loads our dinner plates full of scrumptious food, heaping servings we struggle to finish – mounds of potato and veggies, thick slabs of meat smothered in gravy, gourmet desserts. Dad pleads with Grandpa to give us smaller servings, but to no avail. Grandpa is very set in his strong opinions, an example of what we nowadays would call a 'fixed mindset'. Whenever we pulled up at Grandpa Kilb's house, fear swept over me. His domineering attitude put fear into my heart. Whatever Grandpa says in his household is the iron-clad rule – no questions, no deviations – not even by his son.

'Grandpa, they're here.'

Grandpa storms to the door before Mom and Dad could enter.

'You're not entering my household with a bastard infant who's been abandoned by her mother.'

'Dad, if that's your decision, you should know I will never enter your home again. You either accept our newly adopted daughter as a member of the family, or you can say goodbye to me.'

I'm stunned by Dad's fearless declaration – it's either an adopted daughter, or himself.

Grandpa looks like he's deep in thought as he struggles to blurt out his decision. 'Come in and bring that child with you.'

Brian and I are elated to have a baby sister. We sit on the living room floor, admiring our new family member. She grasps my finger – I can't believe how small her hand is, perfectly formed – an exact replica of mine. Her toes curl up as I tickle the soles of her tiny feet. She breaks a baby smile as I blow raspberries on her tummy. It's hard to believe our Creator can create such a perfect human so small.

Our new sister is called Loral – 'LOR' for Lorna, and 'AL' for Albert. Since that prickly day, Loral has been a loved and cherished member of our family. Loral, Brian and I form an immediate bond – relationships that last a couple of decades before being tragically broken. To this day, Loral and I are extremely close.

Standing Up for Our Values

The bell rings. It's recess time. I throw my pencil onto my desk and scramble to be first out of my grade five classroom. As I burst through the door into the schoolyard, I feel so alive, so free. It's Friday afternoon, almost the weekend. When people ask what subjects I like best, it takes no hesitation to say recess and gym. I love being active, perhaps because I feel confident. I have come to recognise diverse gifts within my classmates, but my Creator has blessed me with the gift of athleticism. The schoolyard is my domain. It's time for me to exercise my gifts as I encourage my friends to try, to learn from their errors, to improve through engagement – but most of all, to have fun.

Today we're playing a pickup game of softball. Most of the kids on the field are my classmates – girls and boys. For me, winning is not that important since we're having so much fun. Susan, an awesome athlete, is at bat. She's not only one of the best athletes in the school but also one of the cutest. You guessed it – I do have a crush on her. As I look at her crouching at home plate, ready to hit, her striking, athletic body and eye-catching face send shivers throughout my body. I get a warm feeling in my tummy. Although I'm one of the infielders, I secretly hope she gets a great hit. Maybe even a double so I can talk to her as she reaches second.

Susan's hit comes screaming down between myself and my friend Tommy at shortstop. Tommy stabs at the grounder, snags the ball and fires it to first. Nick makes the catch and Susan is clearly out. Instead of stepping aside with Susan sprinting towards first, Nick lowers his stance and viciously checks her to the ground. Susan, obviously injured, starts to stand. She's not the type to lie there seeking pity. She's a confident gal who can fight her own fights. Nick knocks her down again. Although injured, Susan musters the courage and energy to get to her knees. Anger wells up within me. My gut churns. My brain turns from joy to rage. I'm furious with Nick's brutal assault. How could he treat a classmate like that? My emotions rise even higher as I look at Susan fighting back the tears as she struggles to stand.

I sprint towards first base and flatten Nick with an amazing tackle. I'm furious.

'What's the matter with you? It's only a game. Why are you picking on someone half your size?'

As I glare down at Nick on the ground, I suddenly realise what I've just done. Nick's the biggest bully in our school. He towers over me and outweighs me by at least ten kilos. His face seems statue-like with a permanent scowl. I'm certainly no match for him. Anger overtook me as I watched him bully Susan. My brain told me, *You've got do something.* Adrenaline filled my body as I jumped into action without putting much thought into what I was doing. Perhaps my reaction was intuitive from my dad – stand up for the underdog and women? Perhaps my crush on Susan had something to do with it?

Nick, lying there on the ground in front of the whole school, is embarrassed. He starts to get up. I realise I'm in trouble. *What should I do? I'm about to take a shellacking. Should I turn and run? No, I can't. I've stood up for what I believe. I can't turn my back on that now.*

Nick grabs me by the shirt and throws me to the ground. He jumps on top of me as I lie on my back. I try to roll out of danger. I'm pinned. His body's so heavy I find it hard to breathe. I cover my face as I prepare for his fists. I'm terrified!

'Hey, stop fighting.'

It's the stern voice of our scary grade seven teacher, Mr. Blackwell. Saved. I never realised the value of a schoolyard supervisor until this moment. If Mr. Blackwell had not shown up in the nick of time, I know my body would have been wearing the wounds of a winless schoolyard scrap.

Mr. Blackwell comes over and pulls Nick off.

'Get back into your classrooms. Now.'

No consequences for fighting? I'm not sure why, except perhaps he had seen the action leading up to the battle and was sympathetic, or he didn't want me to get beaten to a pulp since I'm the best player on his school hockey team. As Nick and I head back to our desks, I glance back to see Susan standing at first, eyes fixed on me, with a smile on her face. Nick glares at me with one of those evil grins, the kind that bullying giants make in storybooks, and challenges me.

'Hey Kilb, let's see how brave you really are. I'll meet you in Dickson Park tomorrow at 10 AM. Be there!'

***

I lie in bed and quiver at the thought of what I've got myself into. Nick's the meanest and toughest kid in our school. I know there will be no Mr. Blackwell to save my butt tomorrow. Showing up will spell disaster for me, but do I really have a choice? Nick will be there tomorrow with lots of his school pals. If I don't show up, I'll lose respect since it was me who started this spat. If I really believe that standing up for my values is important, I've got to show up.

After a night of terrifying dreams – dreams of cheering classmates, of getting pounded, blood streaming from my nose as I lie on my back, I head into Dickson Park with trepidation. Am I scared? You bet. I know how tough Nick is. I understand how badly he wants to whoop my butt for embarrassing him in front of his school buddies. I've got to go and try to protect myself.

It's not hard to find Nick. There's a throng of about two dozen schoolmates gathered to see if I have the guts to show up.

'Hi, Nick. I'm here. What now?'

If looks could kill, I know I would be dead right now. My big concern is that it's going to be more than his looks that will lay harm to me today.

I find myself standing on a patch of gravel, eyeballing Nick's fuming face, veins popping out of his neck, his fists clenched. Perhaps if Nick charges me, he might slide and lose his footing.

My hunch is correct. He charges like a crazed bull. I jump to my left at the last moment. He loses his footing. Down he goes onto his back. I jump on him quickly, straddling him, trying to grab his flailing arms. I can't get them. He reaches up, grabs me in a headlock. In a flash, he throws me into the gravel. My face slams into the ground. His knee digs into my back. His headlock grinds my face into the gravel. The smell of the dry gravel fills my bloodied nostrils. His muscled arm clamps down around my neck.

'I can't breathe, Nick!'

'Do you give up, Kilb?'

'Yes. You win, Nick.'

It's over ever so quickly. I stand up to hear loud cheering from the encircling crowd. *Are they cheering for Nick's victory, or for me being brave enough to show up?* I'm not sure.

No, I did not win the fight, but I do feel like a winner. I walk home with blood dripping from my nose, my face scarred with gravel, but feeling proud. Proud I was willing to give my best, regardless of the odds. Proud I had the courage to face my fears, stand up for my beliefs and values.

REFLECTIONS. Dancing with Fear

We should be thankful for fear. It's part of the human condition. It's a normal human emotion. It's necessary for survival. We must learn to dance with fear, just like the bull fighter dances with the bull. Dancing with fear is a skill. It must be learned and practiced. Start small, accept it, own it, reframe it, get more reps and grow.

Our comfort zone urges us not to be uncomfortable. We must understand that fear will be omnipresent, but we can control it. We need to embrace fear and remember the rewards of throwing away the leash that hinders learning who we are. We've got to decide whether the fear is an opportunity to lean in and grow, or step away. We must be focused and ready to learn, utilising skills we know will help us find our way through the storm. I don't want fear to rob me of a great learning experience. Rather, let moving from fear to fearlessness transform us into more courageous, stronger, confident people – the best version of ourselves, who we really want to be – experiencing the joys of childhood.

I know Dad would have been proud of me for standing up for Susan, although I never told him that story. He would have been even more proud of me for showing up at Dickson Park on that Saturday morning, ready to get the stuffing knocked out of me. Standing up for our beliefs and never backing out of a commitment were two values held in high esteem within our family and ones that I've tried to carry forward into our present-day family.

Accepting fear is part of laying the foundation for a rich and rewarding life. It's that past that dictates the now for us. The challenge as parents is how do we encourage our children to be willing to step into the unknown, the adventurous, those frightening activities? We must find the balance between courage and fear.

There's no substitute for an exemplary parental lifestyle – a willingness to put ourselves in unfamiliar environments. It's not through our words nearly as much as through our bold actions and decisions that leave an impression. Our adventurous spirit is contagious, especially when our children witness or share in the rewards reaped by putting ourselves out there. The greatest gift we as parents can give our children is the gift of time, sharing experiences one on one or as a family! It's in those shared experiences that we are shaped into who we are today.

As a pre-schooler, walking up to a total stranger and following Dad's command was intimidating. It was a frightening and uncomfortable challenge. Dad insisted I do it every time I greeted someone.

'Son, did you look our guest in the eye? Did you give him a firm handshake? I don't want any wet noodle shakes.'

***

At this young age, Dad also encouraged me to engage in 'show and tell' right in our home. He brought home a Meccano set for me to build. When his friends came over, following my handshake, Dad would look at me.

'Son, show our friends what you built today.'

It's these small challenges that helped me at this young age to step out of my comfort zone – to face fear head-on in the safety of our home with the support of my parents. It gradually became a natural habit.

Dad valued my behaviour – his face spoke of the pride that was in his heart. He rewarded my actions whether I failed or succeeded. I believe this is a major reason I'm willing to step into the uncomfortable even to this age. He didn't punish my failures. As he shared his own failures, he taught me that failure is inevitable, that it's the most valuable tool I have in helping me grow.

***

I still remember the first day my dad took me to shower.

'Dad, I can't put my head under that water – I won't be able to breathe.'
'Son, watch me. Just keep your mouth open, and you'll be fine.'
'Dad, I can't open my mouth underwater. I'm scared!'

'Here. Hold my hand and follow me.'

Yes, with Dad's support and validation, I stepped in, mouth wide open, still breathing.

Parents, I implore you to find ways to challenge your children at this early age to form the habit of facing challenges – seizing opportunities to allow growth; grappling with outbursts of fear and unexpected waves of emotion; embracing moderate stomach-churning; and stepping back and allowing them to make their own decisions even if it results in failure. Let's not cheat our children of this rich experiential learning! Let's give them the tools that will help them face anxiety and challenges.

What a lesson to learn at this early age, that fear is meant to be faced head-on rather than avoided. The most important thing with fear is its usefulness, what we do with it and how we can use it. Fear alerts us. Then we have a decision to make – step in or step away. Our ability to make the right decision will assist us throughout life. After eight decades of facing uncomfortable challenges, I'm convinced dancing with fear is the pathway to grow – of stepping in, of utilising the fear to help come out the other side successful, of building confidence and self-esteem.

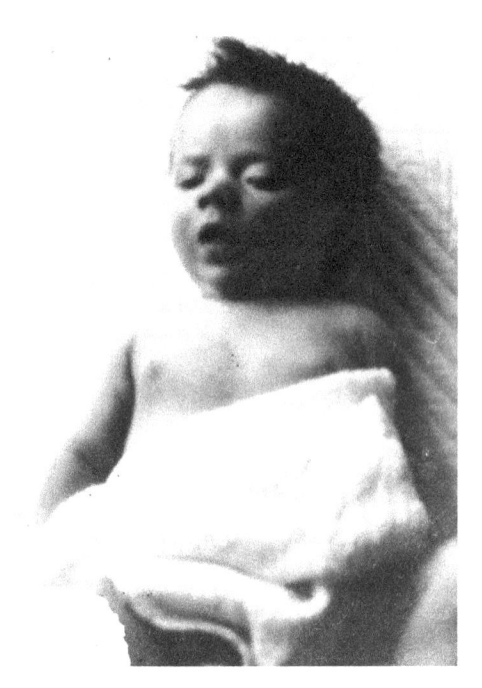

BRAD KILB

Brad, Preschool Years

2. Explore and Discover

'Only those who risk going too far can possibly find out how far one can go.' – **T. S. ELIOT**

The Foundation of Who I Am Today

Looking back at my childhood adventures, I realise they are the foundation of who I am today – mischievous tricks, learning to stand up for my core values, lessons absorbed from my parents, and foolhardy escapades which could've ended my life – the things I loved as a kid make me brave today.

My early years were spent in Galt, Ontario, now known as Cambridge. Our first home was a modest two-bedroom apartment on the top floor of a two-story duplex. Our corner lot was ideally located on the corner of James and Park, the ultimate spot for my exploratory spirit. Just two blocks away was the Grand River, my fantasy playground – casting for catfish and carp; snatching snapping turtles before suffering with fingers in their jaws; soaring through the air on our rope swing before releasing my grip and plunging into my favourite swimming hole; floating downriver on our homemade rafts, Huck Finn style; and jumping on the melting winter ice until it collapsed into the frigid river waters, sometimes marooned on that floating ice block.

The Canadian Pacific Railroad snaked by a couple of blocks north – the ideal spot to carefully place pennies on the tracks as we heard the looming steam engine – our flattened coins no longer legal tender.

Directly across the street was Dickson Park, a large municipal park host to numerous grand-scale events – fall fairs, agricultural showcases, semi-pro baseball, football and soccer games, cricket matches, track and field competitions, shinny hockey on an outdoor rink and horse racing on a 1.6-kilometre dirt track. Our home was surrounded by these spectacular playground locations where I spent all my free time.

I'm truly thankful my parents not only allowed but encouraged me to jump into adventures never before explored. Unfortunately, I believe many parents today put a stop to their children's curiosity. I'm a firm believer that learning to manage risk is much more important than growing up avoiding risk. I want my own children and grandchildren to be street-smart – know when to back away and when to jump in. The fear of exploration can shut down so many amazing growth experiences and feelings of euphoria. I don't want to rob my children of those opportunities. I want to help them learn to dance with fear, engage with their cravings and explore things that spark both fear and growth. It's that past that dictates the now.

My Traumatic First Day

It's hard to contain my excitement. I'm out of bed at six; jump into the new outfit Mom has set out for me; bound into the kitchen for an early breakfast; chatter incessantly about my new adventure. Today is my first day at Galt Dickson Elementary school. I've been waiting for this day for months. With my birthday in December, Mom and Dad decided to keep me back and start school at age six. I'm ready. I so crave my first day in school. I can hardly wait to get out the door. Dad is a schoolteacher, so I've heard lots of stories about school. I'm excited about making new friends, learning new things and playing sports with my schoolmates. I'm anxious to see who my teacher will be.

Mom and I walk across the schoolyard towards the front door. I love sports and the various sport facilities make me feel at home – the baseball diamond, soccer field and barren hockey rink. The large sandstone school looks so big, reminding me of photos of ancient castles I've seen. Mom pulls hard on the heavy wooden door, and we enter. Now inside, my emotions change from excitement to fear. Will my teacher be friendly? Who will be my classmates? Will I do well in school? Dad always tells me that

I must face my fears if I wish to overcome them, so I proceed towards my classroom, gripping Mom's hand.

Mom introduces me to my teacher, Miss Appleby. She reaches down and shakes my hand with a firm grip. Dad always told me to make sure I shake hands with a firm grip. He says it tells a lot about me. Miss Appleby's hand engulfs mine – it sends a message of strength and care. She seems so tall as I look up into her face. I can see her warm smile outlined by her bright red lipstick. Her blue eyes send a warm welcome. Her blonde hair flows down over her shoulders. Her pretty blue dress looks so new – I think she bought a brand-new outfit for our first day, just like me. She asks my name as Mom stays silent and looks down at me.

'Bradley Kilb, but I like Brad.' *I'm not sure how I got it out.*

'My name's Miss Appleby. I'll be your teacher this year, and we're going to have lots of fun together.'

Wow, a teacher who likes to have fun. I feel better already. Mom exits, and Miss Appleby leads me over to a spot on the floor.

'Brad, you sit here while the rest of the class comes in.'

I can't believe it as I sit down. *She even calls me Brad.* Classmates trickle in and join me on the floor in a circle. A bell rings. *I suppose it means the start of the school day. I wish I could slow my heart rate down.*

One of the last students to arrive walks to find his spot in the circle. Suddenly, he stops directly behind me. I feel his fingers running through my hair.

'Hey, look at this. He's got one spot of curly hair right here on the top of his head.'

'Please sit down,' Miss Appleby responds.

'But this part of his hair is so weird. Look at it. It's all curly.'

I can feel my face turning bright red with embarrassment. I fight back the tears as I blink ever so rapidly. *How can this happen? My first day in school and one of my classmates thinks I'm weird. Should I cover my curly spot with my hands so no one else can see? Should I run out of the room and catch Mom before it's too late?*

Miss Appleby comes over and looks down at my hair. *Oh no, now even my favourite teacher knows I'm weird.* I can feel her stroking my hair.

'Hey class, come over here and have a look at this.'

My classmates gather round to stare and feel my hair that resembles steel wool. I can't believe it. My teacher's already making a spectacle of me. *I'm done. I thought I'd love school, but now I know I will hate it.*

'Can you believe it? We have a special student. I've never seen this before. Brad has special hair. We are so lucky to have him in our class. I wonder how he got that special spot of hair?'

Miss Appleby, you're my hero.

My Dairy Bull Teacher

'Bradley, hurry up or you'll be late for your first day with those bulls,' Mom warns.

I slip on my rubber boots, ready to head out the door. My swag is as close to a farmer's son as this city-kid can get. I've not washed my jeans for a couple of weeks. My red plaid shirt hangs over my belt. I've borrowed Dad's leather gloves, worn through in a couple of places. My John Deere cap is pulled down over my forehead as I try to look brave and confident. It's the annual agricultural fall fair across the street in Dickson Park, and I've landed a job. Last year I worked with teamsters preparing their heavy horses for the show – 2,000-pound Clydesdales hooked onto six-horse hitches, pulling massive wooden-wheeled wagons. With my small frame barely able to reach the shoulder of those giant Clydes, I felt so powerful standing on a ladder, braiding their manes and tails. Those docile giants didn't flinch as I combed out the feather flowing over their hooves. This year I'm hired by a dairy farmer to polish the horns of his showcase Holstein bulls before they enter the show ring. I fully realise these dairy bulls are much more dangerous than beef breeds.

I approach my first bull, Theodore, with fear and place my hand on his flank. His massive body fills the narrow stall. I inch my way up the side of Theo. *I swear this beast is purposely leaning against me, pressing me into the wooden slats of the stall.* I can feel his ribs deflate and expand as he breathes. Nearing his head, I see Theo roll his eye back at me, his head swinging so close to me with those long, pointed horns. My claustrophobic reaction changes to fear. I can't stop. I must keep moving forward. Here I am, pinned in the most precarious position – trapped between those perilous horns and the side of the stall. There's no escape.

'Hi, Theo. It's me. I'm here to make you beautiful. Please be my friend.'

I see his eye staring at me, but I'm unable to read his mind. *What's he thinking? Am I friend or foe? Does he like to have his horns polished?* If only I can reach under his neck and grasp his nose ring, I'll have control. My arm feels as if it's shrunk as I strain to reach his nose – my head wavering closer to those horns. This is not a spot a city boy usually finds himself. Finally, I'm able to stretch enough to reach Theo's nose and I feel in control with his nose ring in my hand. It's now time to take out my ebony paper and oil and start polishing his horns.

Theodore is ready for the ring – hooves polished, tail combed out, hide brushed, tail shiny with hair spray, and those beautiful horns gleaming in the sun. Theodore stands tall – proud and ready for the show ring. It's time for me to start again with the second of four bulls. It's another day of victory – I've honoured my commitment, no matter how frightening it may seem – facing my fears and learning to overcome them.

REFLECTIONS. Explore and Discover

I'm so thankful I grew up in an era when children were encouraged to explore, take risks, and discover the world on our own terms – my parents urged me to get out and probe. The terms 'helicopter', 'lawnmower', 'bulldozing' and 'snowplough parents' did not exist back then.

'Investigate what you're curious about, Brad – step into it,' Dad challenged.

'Brad, learn what it means to live in the moment – to live your life with joy and exploration. These moments are the foundation of the rest of your life,' Mom explained.

As I look back on those early years, I'm convinced my values were formed by my parents as they encouraged me to explore my interests. They consistently set the example, not only with words – but by 'walking the talk'. The respect they displayed for every human being was the norm. I watched Dad as he refused to stand silent and watch abuse. His respect for the underdog and women never faltered. He always stood up and became accountable. His courage was remarkable.

My father was a portal into what education really means, not just learning from books but a much broader aspect. From Dad, I acquired a love of the outdoors, and from Mom, I got the true meaning of love.

Unfortunately, we live in a world that sensationalises danger at every turn. As a result, our 'safety first' fixation compels us to do all we can to protect our children as we insulate them from healthy, risk-taking behaviour. As Seneca stated many centuries ago, 'We suffer more often in imagination than in reality.' We hold our children back from flourishing, gaining independence and becoming the leaders they have the potential to be. As risk is removed from the lives of the children with whom I work, I notice an increased level of conceit and a reduced level of self-esteem.

'Dad, why did you seek those reckless adventures when you were a kid? Were you born a gambler? Were you born a crazy risk-taker? Did you grow into an adrenalin junkie? Is adventure your drug? Tell me about your childhood,' my ten-year-old son Bryn asked.

Creating this framework of risk management leads our kids to ask, 'Why not try this? What do I have to gain? What do I have to lose? What's the harm of at least attempting this?' The fear of the unknown becomes diminished when our kids are consistently exposed to new challenges *progressively and sequentially*. They begin to gain comfortable self-reliance in the uncomfortable, even at a young age.

Managing risk is something I spend a great deal of my professional time researching as a university professor training coaches and teachers, as a professional coach and as a mentor to sport coaches and corporate executives. My research is qualitative – discovering truths rather than proving hypotheses. Fear is natural and good. It tells us we must be alert and assess our safe participation. My own experience has shown me that saying 'Yes' when I feel fear will often result in diminishing that same fear.

As I look back at my own youth, I realise my parents and coaches never asked me to do something that I myself didn't believe I could accomplish – attending school on the first day, shaking hands with adults, grooming heavy horses or bulls, winning a faceoff. When I share those expectations, I don't feel the personal pressure that often results in failure.

I remember our daughter Jamey slipping down in her highchair while devouring dinner. Before she plunged to the floor, the crotch strap caught her under the tray. We didn't budge but watched her struggle as she successfully pulled herself back into the chair. We don't believe in 'helicopter parenting'. Dealing with challenges is part of life.

Overcoming challenges is part of laying the foundation for a rich and rewarding life. It's that past that has dictated the now for me. As parents, we must ask – how do we encourage our children to be willing to step into the unknown, the adventurous, the seemingly dangerous activities? In my experience, I have learned there are two important principles we can follow.

Creating a supportive environment, a safety net for our children as they lean in.

Accepting realistic expectations. If our children consent to their own expectations of success, then the challenge will not create pressure.

I'm a storyteller. I love to share my exploits with anyone willing to listen. Many of the lessons I've learned have come through the adventures I've dared to jump into. Those lessons have been burned into my brain, giving me memories I'll carry with me for my entire life. Now, as I record these stories, I feel so grateful that as a young boy I was encouraged to challenge the uncomfortable and unknown as I endeavoured to grow and understand myself. Understanding why I participated in those early risky, and at times dangerous adventures is difficult. As an adventurous adult, I've now come to understand John Kotre, 'This is who I am because this is how I began.'

Brian, Brad   Brad, Brian

Brad, Loral, Brian

3. Learning to Welcome Fear

'Do not live life avoiding risk. Learn to manage risk to experience life to the fullest.' **Brad Kilb**

Ridin' the Rails

After traversing the trestle as it departs the Galt train station, the Canadian Pacific Railway line heads west over the Grand River and Blair Road. We live just one kilometre north of the crossing. Leaving Galt, the laden freight trains sluggishly climb the gradual incline. It's the perfect location to jump aboard.

On this weekend, Pete, John and Steve, three grade-eight friends and I gather on the side of the track and wait for the labouring engine to appear. These guys are my best friends, not only school buddies but also hockey teammates. We hang out together every chance we get. I wouldn't call us a 'gang', but rather adolescent boys running in a pack.

Our excitement peaks as we hear the steam engine chugging up the incline, straining to pull its load. The earth quakes as the 567,000-kilogram locomotive comes into sight. We hide in the bushes just off the tracks to avoid being detected as the steam engine plods along at a crawl. The puffing locomotor reminds me of the children's book, *The Little Engine That Could*.

Once the engine passes with the engineer hanging out the window, we jump into action. It's at this point that I, the leader of our pack, get worried. These pals mean a lot to me. I don't want any harm to come to them. I can't imagine knocking on their parents' door to break the news of their mangled or deceased son!

Most freight cars have metal ladders front and back – the lowest rung about one metre above the railroad ties – an easy jump for us. We know the drill. We've accomplished it before – jumping aboard for a short ride before the train picks up speed. I must decide what type of car we'll board. Boxcars are not suitable since the ladder leads to the roof – not a great place to ride the rails. Even though a boxcar may have open doors, the floor is too high to jump into – a very dangerous choice – one slip, and we're under the rolling steel wheels! Tank and hopper cars have a similar problem: no place to sit while flying down the tracks. Open coal and gravel cars are ideal – a platform with no roof to position ourselves while enjoying the panoramic view, fresh air blowing through our hair.

I examine the massive rumbling cars, my friends anticipating my signal to board. Here they come – our ideal ride – a string of open sand cars. These monsters are five metres tall and forty-nine metres long. The unloaded car weighs enough to squash the life out of us or slice us in half. I shout out the signal over the thunder of the rolling cars.

'Let's go. It's time to board.'

Now I focus on the task at hand. One mistake could result in a catastrophe – an amputation or even death. I don't want to spend the rest of my life limbless. I'm too young to die. But why would we do it? Sometimes we feel the rewards outweigh the fear. To begin with, there's no question in my mind we can all pull this off. We've jumped slow-moving trains before. Each of us is confident. It's another escapade for us to tackle.

We've laid out our game plan. Having all of us use the same ladder wouldn't work. Three of us will be running alongside the train while one friend climbs aboard. By the time the ladder is vacant for our fourth friend, the train will have pulled away, leaving our friend gasping for breath following his long sprint. Two of us will use the front ladder, and two the back.

I sprint alongside the rolling pieces of iron – positioning myself beside the ladder – focusing on my footwork. The stone railbed and railroad ties make sprinting awkward. The uneven footing beckons me to stumble and fall. My brain leaps into action. *Am I sure I can jump aboard this moving car? What if I make one mistake? What if one of my friends is injured?*

It's time to concentrate on boarding. I reach up with one arm and grab one of the higher rungs. My fingers curl around the cold chunk of steel. My grip tightens. I feel like a rag doll dangling on the ladder. I pull with my arm as I jump from the safety of the ground. I plant my foot on the bottom rung. I'm off the ground. I climb quickly so my partner behind me can use the same ladder. I roll over the top of the car and plant both feet inside the box. Safe, all four of us jumping with joy as we realise we've made it. We're ridin' the rails.

What fun. Standing tall with my closest friends, taking in the rural scenery, I feel such a sensation of accomplishment. Our train summits the grade and starts to pick up speed. The faster we go, the more sand pummels our faces as we stand at the front of our car. The sand bites into my face. The air is laden with dust as I breathe in. I squint to keep the particles from collecting in my eyes. I feel the sand accumulating in my freshly shampooed hair. My ears feel like a sandbox as the flying sand settles. The hurtling sand seems as if it'll tear the skin off my hide.

'I've made a mistake. We should have boarded the first sand car,' I confess.

The lead sand car is three ahead. We're now barreling along at high speed. The blast of sand is unbearable. I decide we must move forward.

'How are we supposed to move forward at such a high speed?' asks Steve.

'There's only one way. We've got to cross the couplings between the cars,' I answer.

'That's too dangerous!' retorts Steve. 'A way more dangerous than simply sitting here. One slip and we're done.'

As I stare down at the coupler, I wonder, *Will we be able to cross three of them*? The coupling lurches, jerks and violently moves from side to side. The unstable platform could mean the end of any one of us! *How could I live on if I lost a buddy under this rolling freight?* The span between the cars is two metres, just far enough to prevent us from holding onto both cars at the same time while standing on the coupler.

'We'll go one at a time with the others holding onto our outstretched hand. I'll go first,' I declare.

Why did I say I'd go first with nobody on the forward car to secure my outstretched hand? Maybe we should just remain in this car?

I've got to go first. The terrified expressions on my buddies' faces tells me they will wait to see if I make it before attempting the same ill-advised manoeuvre. My gut is churning. My legs are like slush. My brain is fighting the terror as I try to focus.

I lie flat on the top of the car and swing my legs over. Both hands clamp down on the metal like a vice. I stretch my legs and point my toes as I search for the coupler below. The clickety-clack of the rolling wheels seems to grow louder and more menacing. My feet find the coupler and settle onto it. I'm petrified. I can't let go with one hand to reach for the next car. My white-knuckled grip seems to be my only secure choice as my feet shake violently on the coupler.

'Brad, we've got you. Go for it.' Reassures Pete.

My friends tightly grasp my left hand. I release my right hand and stretch desperately for the next car. My feet thrash around below me. Got it. My right hand firmly grasps the adjacent car. My wingspan reminds me of an eagle gliding with outstretched wings as I glance at the railbed flying by below. I release my grip with my friends and lunge across the gap to clutch the forward car with both hands. I drag myself up and into the car. Safe. With encouraging persuasion, my friends come over one by one. Finally, I stand proudly in the front car with my buddies relishing our triumph, the wind ruffling my sandy hair.

We round a long sweeping bend. I notice the engineer looking back. On spotting us, he shakes his fist. Busted! A few kilometres later, the train slows to a stop. *Is it because the engineer caught us jumping his train?* We'd better deboard and get out of here. We run across a field towards the traffic. We have no idea where we are, except knowing we've been travelling west. Now we've got to get home.

We approach the shoulder of the highway and hesitate to plan. If we all stand on the side of the pavement, thumbs out, it would probably be a deterrent for any driver to pick up all four of us. Two of us will lie low in the ditch, while the other two thumb a ride. A car hits the brakes and pulls onto the shoulder.

'Jump in, boys. Yep, all four of you. Where're you headin'?' the friendly driver asks.

'Thanks for stopping. We're goin' home to Galt,' I gratefully respond.

'You're in luck. I'm headin' home to Galt to my wife and children.'

I usually don't buy a one-way ticket, but today I didn't have to buy any tickets.

The Shortcut

As soon as we move to our acreage just outside Galt, I start looking for shortcuts on my bike. I'm not sure if it's because I'm lazy or efficient. Either way, if I can cut off a few kilometres or a few hills, I'm going to take that easier route.

Coming home along Park Avenue from downtown, Blair Road is on the west side of the Grand River. Most of my activities are on the east side – Galt Collegiate Institute, my high school; Galt Arena, the venue of my hockey games and practices; the national champion and internationally renowned Preston Scout House marching band, where I serve as the drum major. From venues to home is at least five kilometres.

To get home, I have three choices. Descend Water Street into town, turn right onto Park Hill Road, climb the steep hill, turn right and follow Blair Road home. Or, I may choose to climb the hill up Grant Street, or cycle through a field with a dirt path – no hills, no pavement, no streetlights. Travelling through this dark-as-coal field at night is scary. In the pitch black, I fail to see the warning signs of a startled skunk before it sprays. Yep, he drops a stink bomb on me. It's one of the times Mom did not welcome me home!

My third choice is to illegally cross the Canadian Pacific Railway trestle directly west of school, reducing my trip by 3.5 kilometres with no hills. The kilometre-wide train trestle, twenty-five metres above the Grand River, accommodates one train at a time.

Although the dangerous trestle is off-limits, I often cross the catwalk to save time, especially if I'm late for school or my hockey game. I check to see if a train is leaving the station or coming onto the trestle from the west. I head out, hoping to get to the other side before a train appears. When walking, it takes about fifteen minutes to make it from one side to the other. Sometimes, I get caught on the bridge with a train.

The first time is terrifying. Caught halfway across, I see the train barreling down on me. My brain fumbles for a solution – I don't have many! I'm a hockey player, not a track star – sprinting would not help. One metre on the south side of the track is a narrow, rivetted, metal catwalk, less than a half-metre wide – not much security preventing the twenty-five-metre plunge into the Grand River (high diving is not one of my acquired skills). The two metal tracks afford room to lie between, letting the train pass over my prone body – hopefully.

    I decide to lie down on the catwalk, grasping tightly to the rivetted metal until the train passes. I watch it coming like a wall of water in a flash flood, charging down towards me. The engineer shakes his fist as he flies by hanging out the window. I'm not sure if the violent shaking is because of the rumbling train or because of my fear? Finally, what seems like the longest train ever … silence.

***

Crossing the trestle becomes much less frightening as I become more used to the narrow catwalk. One time, with the train proceeding at a slow pace in the same direction as myself, I jump onto the ladder of one of the freight cars and ride across the bridge, jumping off as I reach the other side. No effort on this crossing.

***

My most daring crossing is the day I'm late for my hockey game. I hear the train coming as I ride down Grant Street. I quickly push my bike up the escarpment and onto the bridge. As the slow-moving freight comes along, I sit on my bike with my hockey bag and stick hanging from my handlebars. I reach up and grab the ladder of a freight car with my left hand, my left foot on the bottom step, allowing it to pull me across the bridge as I sit on my bike. Trying to steer my bike across the narrow and uneven catwalk with one hand on my loaded handlebars while grasping the ladder with the other is no easy feat. I make it on time for the start of my hockey game.

Busted

Steve, Jim and I are returning home from our victorious baseball game – gloves in hand and baseball spikes draped over our shoulders. Steve and

Jim are teammates on our hockey and baseball teams. We are neighbours in our new subdivision still under construction, Oakridge Acres in London, Ontario. Our subdivision is a mixture of empty lots, partially completed houses, and finished new homes. A great playground for teenage boys.

'Hey, let's see if we can start that steam shovel and fill in the foundation,' I challenge.

'Brad, we don't know how to drive that piece of machinery. We could all get killed!' Steve warns.

'C'mon Steve,' retorts Jim. 'We just got our drivers' licence, so it couldn't be that hard.'

There's no holding us back. We charge over to the huge steam shovel precariously perched on the edge of the excavated foundation. The operator had spent the day digging the hole for a new build. We lay our baseball gloves and spikes down and climb up onto the track of the machine. The window's been left open, making entrance oh so easy.

'Brad, are you crazy? Get out of that operator seat,' advises Steve.

'I think I'll be able to figure out how to start it.'

'Don't be stupid, Brad,' continues Steve. 'You're sitting on the edge of the foundation. If you start this machine and it's in gear, you'll roll right into that four-metre foundation.'

As I look at the overwhelming number of pedals and levers, I realise that if I roll into the foundation, I'm done. There's no way I would be able to escape this cabin in time.

My mind's too busy trying to figure out how to operate this machine. I ignore my friends' warning. I'm not sure which levers operate the bucket, which lever controls the RPM, and which pedals control the tracks. It's all too complicated for me. I've never operated such a machine.

'Hey, Jim and Steve, I think this button is the starter,' I say.

'Don't push it, whatever you do,' pleads Steve.

I reach down to push what I believe to be the start button. I see my buddies abandon me as they jump off the track. I say a short prayer, hoping the gears are in neutral. The powerful engine turns over. Exhaust plumes from the stack. The sputtering turns into a roar as the Caterpillar engine fills the stillness of the evening.

'I did it. I've got this sucker started,' I claim with delight.

The looks on my friends' faces do not correspond to my exuberant joy. With the steam shovel solidly stationary on the ground above the foundation, Steve and Jim jump back up onto the track.

'Get in here and let's have some fun,' I beckon.

'Are you kidding?' cautions Jim. 'You still haven't figured out how to move back away from the foundation.'

'You can sit in your own coffin, Brad, but I'm not joining you,' warns Steve.

Jim is right. Starting this baby was easy, but I'm not sure the rest of the manoeuvring will be so straightforward. *Where do I start? How do I figure out what lever does what?* There is no room for error – one metre forward, and I'm somersaulting into the foundation.

'As long as I'm in neutral, I can try to figure out how to operate the bucket,' I confidently announce.

I slowly start to engage one lever at a time. I find the manual accelerator and give 'er more gas. I press a lever that moves the arm of the bucket. I keep experimenting until I discover the levers that operate the arm and bucket angles. It's time to start having fun.

We have so much fun scooping the soil from a large pile back into the foundation hole. It's not nearly as precise as my days operating my Tinkertoy steam shovel, but we're gettin' 'er done. Scraping the soil back into the hole is much easier than filling the bucket and dropping it. We take turns and cheer each other on as we get better and better.

With our intense focus on our new-found skill, we fail to notice the black and white speeding towards us on the dusty road.

'Hit the kill button and let's get out of here,' Jim shouts.

We jump down and sprint across the vacant lots. As I look back over my shoulder, I see the cop pull up and stop next to the steam shovel. All that practice stealing bases with our baseball team pays off as we appear to have put in enough distance between ourselves and the cop.

Giggles and smiles quickly change as we see the cop place our gloves and spikes on his patrol car hood.

'Damn. He's got our baseball gloves and spikes,' I declare.

'Now what?' Steve murmurs.

'Let's see if he leaves with our confiscated loot,' Jim answers.

Our wait for the cop's departure is futile. We assume he's used to long stakeouts. We sheepishly walk back after a short hiatus.

'So, boys, I got a call from that house right there that you've been operating this steam shovel. Is that true?' snarls the officer.

'No, sir. We've no idea how to operate a steam shovel,' I falsely claim.

'Well, let me check out your story.'

He strides over to the steam shovel, climbs onto the track, leans across the engine and puts his hand on the exhaust stack.

'Owwwww. Damn!'

The hot stack burns his hand. He jumps down from the track, shaking his scorched hand.

'You sure as hell have been driving this machine. Get into the back seat of my patrol car,' demands the cop.

'Is it OK if we get our gloves and spikes first, sir?' I ask.

'Hurry up.'

We slide into the back seat, gloves and spikes on our laps, not daring to make eye contact with each other.

'If there's any damage to that steam shovel, you're paying for it,' the cop declares as he reaches for his notepad. 'I need your names and addresses.'

I could see him fumbling with a pen and tiny pad between his burned, now blistering fingers. His face winces as he tries to write.

'Brad Kilb, sir.'

'Spell it kid.'

'K – I – L – B.'

'Jim Wettlaufer, officer.'

'How do you spell Wettlaufer?'

'W – E – T – T – L – A – U –'

'How the hell am I supposed to get that name onto this tiny pad?' the angry cop utters.

He'd run out of room to spell Jim's last name. He's not a happy cop. I start to giggle and can't stop. This doesn't make him any happier as he turns and glares at me.

The cop drives each of us home. As my parents open the front door, I can see their shocked looks as I'm accompanied by a cop with his patrol car sitting in our driveway. Dad grounds me for a couple of weeks but congratulates me on the fact there was no damage to the steam shovel.

The Clock Shows Four Seconds

Our high school hockey team at Medway High in London has made it to the city championships against South Collegiate. The championships are a two-game home-and-home series; total goals win the trophy. South has won our league game earlier in the season and is up by two goals as we move to our home arena to play the second game. All odds are on South to walk away with the hardware.

By the end of the first period, we've closed the gap to one goal. By the end of the second, we've tied the overall score. Our fans are going wild in our own barn. In the third, South scores another to take the lead. We battle hard but to no avail. The linesman whistles an icing against South with four seconds remaining. Coach Hoople pulls our goalie and calls a timeout.

'Kilb, take the faceoff. You've got to win the faceoff. Direct the puck towards the net. Everyone up and charge the net on the drop of the puck. No passing. Just knock the puck towards the net. We've only got four seconds!'

I glide slowly into the faceoff circle. The home crowd is in a frenzy. I'm so afraid. This championship depends upon me! I've got to focus on the now – the fundamentals of winning this faceoff. I so want to be the game-changer.

I'm now able to slide into one of the greatest gifts of facing fear – the ability to move into a quiet time. That time when I become completely focused on the moment – the task I must execute. There's nothing else happening in my world – no fans, no opponents, no past or future – full concentration on winning the faceoff.

I shorten my grip on my stick; widen my stance; take my time dropping my stick blade to the ice. As a left-handed shot, I've got the advantage on this side of the net. In one motion, I can knock my opponent's stick out of the way and come back with a shot on net. I want to anticipate the linesman's release of the puck. My eyes focus on the linesman's hand. His wrist flexes. He drops the puck. I make my move. I feel the puck come off my stick towards the net.

I can't see the puck. Where is it? The goal light goes on. The crowd erupts. We've tied the score. We've done it. My teammates mob me. The crowd's chanting, 'Kilb, Kilb, Kilb.' It's one of the most exciting moments of my young athletic career.

We prepare for the sudden death overtime period, exhausted physically and mentally. South scores. My heroics are short-lived and soon forgotten – except by me.

REFLECTIONS. Risk Management or Risk Avoidance

As I look back on my life, I've felt fear so many times in all areas of my life – asking a girl to dance with me at the school dance; hearing that cancer has invaded my body; asking my wife to marry me; raising my children; interviewing for a job; speaking in front of large crowds; being vulnerable as I open up and become exposed; swimming with a shark and whale. Overcoming those fears is a learned skill – it takes guts and practice.

As children, we look at fear in many different ways: some of us not daring to venture there; some inviting engagement, craving the excitement. Having parented six kids myself, I firmly believe that we as parents must introduce our kids to challenges and adventures. There is no question that all of us have a different level of engagement in fearful adventures. Stepping into the unknown for some of us is inviting and exciting. While to others, it's terrifying. My only suggestion is that we introduce our kids to risk in an *intentional, progressive and sequential manner, with support.*

I feel so grateful that as a young boy, I was encouraged to jump into the uncomfortable and unpredictable as I endeavoured to grow and understand myself – not always on my own volition, but often because of my parents. From the get-go, Mom and Dad wanted me to experience life to the fullest. They made learning 'curiosity skills' a priority. Most children need to be taught how to be more curious. Mom and Dad were willing to accept the truth that if they allowed risky play, they must live with the consequences – bleeding lips, black eyes, cuts, abrasions, broken bones and perhaps a fragile psyche.

Both my parents had that exploratory mindset. They pushed the limits as they took on challenges to travel and grow – Dad serving as the principal of the Kumasi Technical Institute in Ghana and the Singapore Institute of Technology; flying to Germany to pick up their VW Westfalia, which became their home on wheels for eighteen months throughout Europe, North Africa and the Middle East. The thrill of travelling excited Mom as she fully endorsed these adventures with Dad.

How do we decide where to draw the line? When do we sweep in to remove the need for our kids to navigate hardships and solve problems on their own? When do we step back and allow them to fail – to reflect, learn and try again? Is it our responsibility to deprive our children of gaining insights into the world and understanding oneself – something so innate to human beings? Without risk, there is no gain. For me, I draw the line if my assessment could dictate life-impacting injury, emotion or death. It's then that I decide to say 'No', or join my children, supporting and teaching them how to do it safely. Messing up can result in consequences allowing us to grow wiser and stronger – developing the self-discipline to make better choices. Natural consequences can be some of life's greatest teachers.

Challenges, whether physical or psychological, can be terrifying. Actions we would sometimes rather walk away from than lean in to. Each one of us has a different level of tackling the unknown. It's important to realise that every one of us has a different tolerance for risk. As a result, we could lose some very impactful learning moments. Mom and Dad encouraged me to step into that arena, to become engaged, to dance with fear, to analyse success and failure and to make wise decisions. No matter how small the challenge, the positive outcomes can be immense, even if we fail. Learning from our setbacks is perhaps the most practical way we can grow. Our willingness to be vulnerable, to participate even when we're not sure of the outcome, will lead to growth. Unfortunately, I believe so many parents today put a stop to their children's curiosity.

Failure is a natural part of navigating life. There is no shame in failure, only shame in not learning from it. The best entrepreneurs fail all the time. The difference is that they leverage those lessons and learn to set themselves up for a better chance of success the next time around. Turmoil, chaos and failure should be embraced as a way to learn, which in turn reinforces a risk-taking mentality.

Children need help in understanding that in anxious situations, they must focus on the fundamental steps leading to success rather than stressing about the outcome. It's what forces us to focus intensely – to live in the moment as we fully concentrate on what we must do. Bravery is a willingness to act courageously in the face of adversity – stepping into fear with something on the line – a willingness to risk your neck.

Feeling fear is OK. I believe being brave isn't living in the absence of fear – it's facing fear head-on. With my athletes and children, I will often script fearful scenarios for them to fight through. Perhaps some of those scenarios will result in failure. No problem. Then it's the debriefing that enables us to let children try without parental interference – to learn from mistakes in a safe and supportive environment, to fill the toolbox with street-smart skills.

So many of the lessons I've learned in my life have come through the adventures I've dared to jump into. Opportunities I've seized that may never come by again. Lessons that explore my gifts, exemplify my priorities, illuminate my values and consolidate my goals.

Mom and Dad taught me that rejection, failure and unfairness are part of life. They never failed to give me the guidance and support I needed to deal with those setbacks so I could gain confidence in my ability to handle life's inevitable hardships. Brene Brown is a research professor and best-selling author who's spent the past two decades studying courage, vulnerability, shame and empathy; believing that we must walk through vulnerability to get to courage. Her motto, 'Courage over comfort', speaks so deeply to me as we lost our twenty-eight-year-old son Brett and forty-six-year-old son Brad Jr.

Train Trestle Shortcut

4. Leaning Into the Unknown

'Until you step into the unknown, you don't know what you're made of.' **Roy Bennett**

African Adventures

We arrive in Kumasi, Ghana, in August 1961. My father's been hired by the Canadian International Development Agency to be the principal of the Kumasi Technical Institute. His role is to guide the Institute and at the same time, mentor a local Ghanaian who will step into the principalship the following year. Our whole family is living in Ghana. Mom is the housekeeper, my younger brother Brian is studying at art college, my young sister Loral is attending high school and I'm enrolled in the Bachelor of Science degree programme at the University of Ghana, an affiliate of the University of London, in the capital city of Accra. Accra is 250 kilometres south of Kumasi on the coast of the Atlantic, so I moved into residence and made it my new home.

As a science student at the University of Ghana, I'm hoping to utilise my Bachelor of Science degree for entrance into dentistry. I've decided dentistry would be a good profession for myself – able to make lots of money, and control my hours and time off, which would allow me to pursue my real love – outdoor adventures.

As I climb into bed for my first night in this new country, I check to make sure everything is set for the night. I'm still bathed in sweat as I crawl under my single sheet. The ceiling fan is rotating, slightly circling, circulating the stifling heat. A small beam of light allows me to follow the geckos scurrying up and down my bedroom walls and across the ceiling above my bed. My mosquito netting encircles my body, protecting my skin from those malaria-carrying mosquitoes. Lying there, tired but alert, I hear the constant drumming from a nearby village. The drums seem so close. *What do they mean? Do they drum every night? How safe am I in this strange new country?* Finally, exhaustion takes over, and I fall into a deep sleep on my first night in West Africa.

Living in a country which is so different can fill our imagination with fear. It's a prime example of *perceived fear*, which can be totally unwarranted. In this case, my fear is absolutely unfounded. My year in Ghana is filled with exhilarating but safe escapades. It's often the unknown which plants fear in our brain!

Speed Bump

I pull into our driveway on my motorcycle, and my sister Loral runs out to greet me with big hugs. It's so good to be home again and see her smiling face. I invite Loral to jump on behind me, and we'll head out. Loral teases, 'You remind me of Steve McQueen in *The Great Escape*.'

We're cruising along a country road when I spot it ahead. My heart jumps. I can't figure out what it is! I'm hurtling down the road too fast to stop. It's so big that it stretches across the entire width of the pavement. *Can I manage to manoeuvre over it, or do I hit the ditch?*

'Lift your feet. We're about to hit it,' I shout.

I secure my grip on my handlebars. Lift my feet off the pegs. Thump, thump. We hit it. The bike shudders as if we've gone over a speedbump. As we look back, I realise we've run over a boa constrictor slithering across the roadway. Had we stopped beside that predator, we may have ended up as lunch for that huge snake!

No Headlight – No Problem

My right hand shadows the brake lever on my Matchless 650cc motorcycle. To get back and forth from my university campus in Accra to see my family living in Kumasi, I ride my powerful bike. After one of my trips home, I'm heading back 250 kilometres south to Accra in the pitch black with a burned-out headlight. I've got to get back on campus for my Monday morning lectures. My dangerous trip will take about five hours of white-knuckled riding.

My brain is on full alert! I'm terrified! With no headlight, I'm afraid I'm going to slam into a pedestrian, horse-drawn cart, cow or parked vehicle! My eyes strain to spot shadowy objects on the road. Maybe I should pull over and continue tomorrow, but I have no place to stay, and I'd be late for morning classes. It's so dangerous. I pull over to conceive a plan.

I decide to wait for a lorry, accelerate quickly, and tuck in behind it. As I fly down the highway, three metres behind the lorry, I'm petrified the driver is going to slam on his brakes. *Do I have time to stop before slamming into the back end of the lorry, decapitating myself on the truck bed?* I decide to pull back a bit so I have enough time to react to the lorry's brake lights.

As I go through villages, I quickly realize that people dart across directly behind the truck, not realising I'm there without lights. I swerve to barely miss a couple of kids as they dash across the road! My mind struggles. *Do I stay back with the possibility of hitting someone crossing the highway – or do I pull in close to the truck and fear ramming into the lorry deck?* I decide to close the gap, stay close to the lorry, riding right on his tail with my hand on the brake. It's one of the most terrifying rides I've ever been on.

Juju Priest Caring for His Ancestors

'Don't let that deadly croc get behind you,' warns the Juju priest.

A Ghanaian friend of mine shares a tale of a Juju priest who claims to have a special relationship with crocodiles. My curiosity takes me to his northern Ghanaian village near Tamale. I've got to investigate this religious practice.

I'm introduced. His striking headdress of chicken feathers give the image of a much taller witch doctor. The blood-coloured paint across his forehead and encircling his eyes give me an eerie feeling. Large white croc teeth protrude from his black nostrils, and a necklace of razor-sharp croc teeth encircles his neck. His crimson loincloth sways below his waist, adorned with feathers and teeth.

We proceed to a chicken coop where he stuffs about a dozen live chickens into his croc-hide bag. The chickens squawk incessantly as we head down to a small lake bordering the village. At the lakeshore, he reaches into his bag, grabs a chicken by the legs, pulls it out and shakes it mercilessly. The hen clucks ever so loudly. Perhaps it understands its doom.

The calm surface of the lake is broken by about two dozen crocodiles. I see only their eyes, ears and snouts. The crocs swim stealthily towards us. I'm feeling more uneasy. I'm realising the chicken will probably be a snack for these crocs, but I'm not keen on being on the menu. The Juju priest is standing about four metres from the edge of the water on the flat, sandy beach. I'm having trouble trying to understand why he displays no fear. My brain is overflowing with fear! I position myself behind the priest, not really convinced I'm out of danger.

The priest continues to shake the chicken viciously as the crocs reach the shore and waddle ever so close to us. I've heard that crocs are very fast on land over a short distance, but I dismiss that thought as I attempt to put on a brave face. The crocs open their cavernous jaws. I stare down at the rugged, armoured, green-skinned skulls. The sight of their eighty jagged teeth terrifies me. Crocodiles are the world's champion chompers, killing with the greatest bite force ever measured for living animals. With a slight backswing, the priest hurls the squawking chicken down the throat of the closest croc. The trap closes with one chomp. He swallows, then turns and disappears under the water. One less chicken, one happy croc, one pleased priest, one terrified Canuck.

As the crocs form a semi-circle before us, I make sure none of them get behind me, trapping me in between them and the water. I feel a little safer in that position, although I'm not sure I would be able to escape if any of them decides to attack. The experience is hair-raising as the priest pulls out chicken after chicken and throws it into the gaping mouth of another croc. I feel very relieved as he pulls the last chicken out of his bag.

As we slowly back away from the lake, I ask the priest why he feeds the crocs.

'When a villager dies, they come back reincarnated as a crocodile in this sacred lake. It's my duty to feed them daily.'

'Has any priest has ever been killed by the crocodiles?'

'Oh yes, priests have been killed in the past. We villagers know those priests were evil. It was the crocs' duty to remove those priests from their privileged position.'

'How can you be sure your ancestors knew those priests were evil?'

'That's easy. You see, the crocs never eat the evil priests. They just drown him and leave their bodies to rot.'

Makes sense to me, although I'm extremely happy the priest I joined today was not evil. Watching chickens being eaten alive was enough of a rush for me. Evil or not, I was not ready to see a Juju priest drowned by his ancestors.

Juju Priest's Power Over Crocodiles

News reaches me that a village is being harassed by a three-metre croc. The villagers believe this crocodile lives on a small island about 100 metres into the lake bordering their village. The croc has come into the village on numerous occasions for a canine snack. The villagers are worried the croc may come and take one of their children and have hired a Juju priest to kill the croc today (not my Juju friend, the gourmet priest). I'm curious to see how this Juju priest slays a dog-killing croc.

As I arrive, it's not difficult to locate the priest. A large crowd has gathered at the lakeside and are chatting in excited voices. I make my way through the crowd, and there he is, wearing a necklace of crocodile teeth and a fancy helmet made of cow horns and pigskin. His confidence and determination indicate this is not his first crocodile rodeo. He does carry a gun, which gives me some hope.

He chants something I don't understand, throws some incense into the air and proceeds to wade into the lake. I hear that his plan is to wade out to the island, find the giant croc and shoot it.

'Aren't you afraid the croc will get the priest before he gets to the island?' I ask the mother as she clutches her child.

'Not a chance. He has supernatural powers over crocodiles - they're unable to harm him.'

I watch in disbelief as the priest wades out to the island in chest-deep water, gun held high over his head. I'm expecting any moment to see the priest be devoured. There's no way he would be able to utilise his gun if attacked! My brain tells me this is a classic case of predator becoming prey. The crowd is silent, with only a few murmurs as they watch. I don't whimper a word. I realise I'm uncontrollably shaking with terror. I anticipate observing my first-ever infamous croc death roll.

Much to my disbelief, the priest ascends from the water upon arriving at the island. I can't believe he's still alive. Maybe there's more to this Juju power than I know. Should I take up this practice while here in Ghana?

After a short time, we see the priest re-enter the water and wade back to the village. I'm flabbergasted. He's made it out to the island and back. My only explanation is that there is no croc in the vicinity, or the croc is full of dog. As I depart from the village, the villagers are depressed, but I'm happy. I've lived another eventful day to see a Juju priest beat the odds and remain alive.

Snake Oil Charlatan

'Brad, why don't you come home with me this weekend. There's going to be a snake charmer in our village.' George's invitation is one I can't turn down.

'Of course. I'd love to go home with you to meet your family.'

George Affori-Atta is one of my closest friends at the University of Ghana. He's not only a close friend, but also my mentor as I acclimatise to the Ghanaian lifestyle. George, of Ashanti ancestry, dresses immaculately and carries a little extra weight, with skin as dark as coal, white eyes bursting from his smiling face. His open and friendly demeanour attracted me during my first week of classes. We spend hours together in the resident dining room and study hall.

After an hour on my motorcycle, we pull into his isolated village along a dusty trail. Before George can dismount, his wife wraps her arms around him and lays a welcoming kiss. The villagers are visibly excited to see George and his Caucasian chauffeur. I turn off the engine, swing my leg over my bike and pull it onto its stand. I notice the children peering out from behind their parents' legs.

'George, it seems like the children are afraid of me?'

'Yes, many of them have never seen a white person before.'

MY LOVE AFFAIR WITH FEAR

As we walk towards George's compound for dinner, some of the children bravely approach and apprehensively touch my white skin. Wow, what an astonishing experience for me. It's inconceivable for me to fathom that a child had never encountered a white person before. The children are so inquisitive and cute. I decide I'd like to take one of them home to Canada with me, but adoption at my young age is not in the books.

Once fed, I can feel the excitement building in the air. The villagers encircle a large bonfire as the drums beat. I'm sure the gathering must be for the snake charmer George alluded to. We pull up a stump and await the arrival of the charmer. I'm not sure what to expect. The only serpent charming I've witnessed is the capturing of a harmless garter snake.

At sunset, the drumming ceases. Silence falls over the crowd. All eyes are glued on the small gap in the crowd. The Juju priest and his assistant make their entrance. He delicately sets his enclosed basket on the ground. The flames dance on his face – a face creased with lines, the map of a hard life. His icy stare pierces the darkness above his white beard. His expression is one of assurance – no fear of what's to transpire.

He extends his arm towards a basket, reaches inside, pulls out a musical instrument and places it on his lips. Music fills the air as he starts to play. Another basket shudders. The lid rises only slightly. Mothers clutch their children and pull them back. Goosebumps arise on my body.

His assistant reaches towards the basket as if to lift the lid. The growing crowd gasps. But no, he seems to want to prolong the act. Again, he grasps the lid, unhurriedly removing it. The terror is palpable. Slowly a cobra rises as if drawn by the tune, rearing upwards out of the basket. Facing the charmer, the venomous cobra spreads his hood. The serpent sways back and forth in unison with the charmer – in time with the music.

The cobra strikes. The charmer leans back. The assistant readies his snake stick. The charmer is unscathed. The serpent retreats. The crowd looks on in disbelief, white eyes wide open. Again and again, the snake strikes but to no avail. It seems as if no harm can befall the charmer. Does he have power over this venomous creature?

Finally, the music stops. The snake retreats into the basket. The lid is secured. The invincible charmer stands, victorious. The crowd erupts in cheers. The show is over.

But no. Now his assistant opens his bag to display numerous bottled potions – various magic spells against snake bites. The mesmerised villagers scurry to purchase the potions. The charmer and his assistant leave the village a little richer – the villagers a little poorer but feeling a little safer.

***

Before joining George on our weekend excursion, I did a little research in our university library. Snake charming is typically an inherited profession. Most would-be charmers learn their know-how at a young age from their fathers. The earliest snake charmers were likely traditional healers by trade, living a nomadic existence, concocting and selling all manner of potions. As part of their training, they learned to treat snake bites and how to handle snakes. Traditionally, snake charmers use snakes they have captured themselves in the wild. This task is not too difficult since cobras tend to be slow movers. The exercise also educates the hunter on how to handle serpents.

During a performance, charmers take several precautions – sitting out of striking range, plugging the fangs with wax, or even removing the fangs. Although snakes are capable of sensing sound, they lack the outer ear that would enable them to hear music. They follow the rhythm of the music. As the snake emerges from the basket, it sees the charmer as a threat, responding to it as if he were a predator – instinctively spreading its hood when threatened.

Personally, I find it inconceivable to believe these entertainers have magical powers or life-saving concoctions – I consider them to fit into the category of charlatans, making a living from ignorant dupes.

***

Although many of my skills and attitudes have been passed on to me by my father, this is one skill I'll forego – venomous snakes fill me with terror. While studying at the University of Ghana, I leave the study hall late into the evening. Both sides of the road are lined with deep, open rain gutters. I'm so afraid I walk down the centre of the road, one step at a time, keeping a sharp eye out for a rare, attacking snake. The most common sub-Saharan snakes include pythons, cobras, African adders and the most venomous – the black mamba. Black mambas, often called the 'coffin snake', can grow to three metres and are among the fastest snakes in the world. Thank God I never have to escape the attack of a black mamba – I never carry a charmer's flute!

About to be Sacrificed?

It's early Saturday evening. The week's classes are finished, and I fire up my motorcycle. I'm heading into Accra to one of my favourite weekend nightclubs.

As I walk into the open-air nightclub, the live band plays the Ghanaian favourite dance music – the Highlife. Highlife is a recreational dance genre celebrating the spirit of independence. If I must say so myself, I'm getting pretty good at this dance genre. Several local girls have tutored me as I've stepped all over their feet. I love the beat.

The club is filled with happy young people – twenty-somethings. I notice a few businessmen scattered throughout the crowd – they stand out like misplaced expatriates. A table of friends beckon for me to join them. I've spent many previous Saturday evenings with them – drinking Star Lager beer, chatting, laughing, backslapping and dancing. My Saturday nights have become a joyful routine with my Ghanaian friends. I greet them in pidgin English.

'How far?'

'How you dey?'

'I dey fine.'

I've become quite fond of a pretty local gal, Afia – tall; long, black-braided hair; a wide smile with white teeth gleaming; sparkling brown eyes; dark skin; shapely, athletic body; joyful, fun-loving; speaks perfect English. She pulls up a chair, and I gladly plunk myself down. The fun continues late into the night as we dance the Highlife under the stars. As I'm ready to say goodbye and head back to the university campus, Afia invites me to spend some time with her.

'Hey. Would you come home with me next weekend? I'd like my family to meet you.'

'Do they live in Accra?'

'No, but we could catch a lorry out to Koforidua, my village.'

'What day would we go?'

'I could meet you on Friday afternoon, and we could come back to Accra on Saturday.'

'Are you sure that would be fine with your family?'

'Oh yes. They've heard all about you, and they're keen to meet you.'

'OK. Why don't I pick you up, and we'll travel together to your village?'

'Great. Let's meet here at the club at 4:00 on Friday.'

She's just so much fun to be with, and I jump at the chance to experience village living with her family.

***

We turn off the pavement and head down a dusty road, following Afia's instruction. I hear her giggling, wind blowing through her braids and a huge smile on her face as we pull into Koforidua. I have no idea how she's let her family know I'm coming, but it's as if I'm a rockstar entering the village. The whole village turns out to greet us. Afia is in her element. She seems so excited and pleased to introduce me to her family with her arm around my waist.

As the chaos settles down and introductions finish, we sit down to a typical Ghanaian dinner – fufu (made by pounding a mixture of boiled cassava and plantain into a soft, sticky paste that looks and feels like white peanut butter), served in a bowl of aromatic and spicy goat soup (with pieces of goat meat and eyes, some meat still sheathed with hairy goat hide). The tang of burning dung fills my nostrils as the smoke drifts my way. We eat with our fingers as the cooking fire smoke stings my eyes. I wash my fufu down with sweet, home-brewed Akpeteshie – the national spirit of Ghana, distilled from palm wine or sugarcane juice, resulting in a drink that is forty percent alcohol.

As the sun sets, Afia leads me to a mud hut with a thatched roof. She pulls back the curtain serving as a door. On the mud floor is a double straw tick covered in clean white bedsheets, a blanket, and clean pillowcases. The space has obviously been prepared for this 'rockstar' who had no idea I'd be spending the night alone with Afia.

'This is where we'll spend the night,' she claims.

'Don't you think we should spend the night in your parents' hut?'

'Oh no. This is where they want us to stay.'

Afraid of insulting her parents and the villagers, I agree.

I climb into bed with my mind racing. I saw the name of Afia's village as we drove in. It seemed to ring a bell. I searched my memory. *Where had I heard this village's name before? I suddenly realise what I'd read in our university library – this village was infamous for human sacrifices!*

My mind swirls, imagining all kinds of horror stories. My logic struggles to take over. *Ghanaians don't still allow human sacrifices. I'm sure there has not been a human sacrifice for decades.* But who would ever know if I were to be sacrificed tonight? There would be no trace of me whatsoever. I had left no word with anyone about where I was going this weekend. I would simply vanish from our globe, no trace of my sacrificial corpse.

How well do I really know Afia? Is she really the pretty girl I dance with every Saturday night? Or is she the bait who delivers Caucasian men to be ceremonially sacrificed (that had never crossed my mind)? What will the sacramental procedure look like? Will I be dismembered, sliced into pieces, or burned alive – not my idea of cremation!

I fight to remember what I had read. *How can I go to sleep? Can I sneak out on my motorcycle without being noticed? Shall I wait until Afia is asleep?*

The night seems to linger on as I thrash about in my sleepless trepidation. I'm so relieved to see the sunshine through my cloth door this morning. As I ride back into Accra with Afia gripping my waist, I thank God for allowing me to live another day.

Sacrifices were particularly common in the Benin Empire, in what is now the Ashanti Region of modern Ghana. Human sacrifice is defined as the act of killing one or more human beings as an offering to a god. Many people in the communities where these practices are still held have accepted them as a part of their tradition and do not report them to the authorities. Perhaps I wasn't considered a worthy sacrifice? Whew!

First Night in Camp

It's my first night in the Nigerian Outward Bound School. We gather in a large tent for orientation and to meet the instructors as they outline their expectations. I look around and realise I'm the only white on the course. Most of the other students are Nigerian Armed Forces officers, compelled to attend as part of their training. As introductions commence, I hear a loud explosion. Scurrying out of the canvas tent, I witness a fireball piercing the night darkness.

I sprint up the embankment, reaching the top, gasping for breath. There it is – a downed single-engine aircraft engulfed in flames. Smoke spirals skyward. The fuselage is severely damaged. The wings and tail are still intact. Two shrieking men are trapped in the cockpit. My brain is spinning as I confront the horror of the crash. *How can this happen on my first night in camp?* Realising the two men are still alive, I grab another student,

'We've got to try to get that pilot out of the fire!'

'Brad, are you sure the plane won't explode?'

'We've got to go in now. He's going to die!'

The stench of heavy smoke fills my lungs. Spotting a window of access to the cockpit, we move in to grab the screaming victim. We pull on the pilot's arms. He doesn't budge. He's belted in with his seat belt. The heat is intense, the smoke blinding. Breathing is tough as I suck in the fumes! I slide my hand down his body to his waist and grope for the seat belt. The blazing flames sear my skin. The impregnated gas smoke burns my eyes.

'I've got it.'

My hand burns as I reach the scorching metal. My fingers grapple for the release – the pain's like grabbing a frying pan on the stove. I pull up on the hot buckle.

'I got it. He's free.'

We pull on his arms. The pilot pops out of the cockpit. We fall back onto the grass, his limp body landing on top of us. I can feel the heat coming from his body. His screams drop to a whimper. We've extricated him from the burning inferno. He's alive.

Witnessing our actions, other students pull out the other victim. We administer first aid to our delirious victim, attending to his raw burns and bleeding. Unexpectedly, one of the instructors appears and instructs us to stop. To our surprise, the two victims stand and join the instructors.

The aircraft crash is the first of many leadership assessments – a mock crash utilising a real airplane body and carefully placed flames, the injuries of the victims professionally applied by a make-up artist. The instructors have been watching our every move and assessing our leadership qualities as we react to this disaster. The next twenty-eight days are filled with similar episodes that repeatedly challenge me. The debrief back in the tent sets the stage for the next four weeks. We will be tested again and again in numerous outdoor activities designed to build strategies for dancing with fear, as our instructors strive to bring out the best in each one of us.

Outward Bound Journal Entries

Today our exercise is to rappel thirty metres down a sheer vertical cliff. As I start over the edge, I'm frightened. I can feel the rope stretch as it scrapes across the rough rocks. I can feel my shoes slip. But I keep going. Little by little, my confidence is gained, and I enjoy the descent immensely. It feels great for me to push myself out perpendicular from the face and feel myself hanging there in space supported only by a rope. I'm doing something I've always dreamt of.

Outward Bound Journal. Feb. 11/62

***

The best part of the day is the late afternoon as we head down to the river to bathe. Relaxing in the dirty brown water is such a relief from the heat. But the warning shouts from our scouts end at all – 'Crocs, Crocs!'

Outward Bound Journal. Feb. 14/62

***

This morning I wake up happy and raring to go. Little do I know, but I'm about to spend my day in hell. Never have I experienced such unbearable circumstances. Never have I been subjected to such extreme heat with absolutely no way of finding relief. Our job is to build a six-kilometre road into this isolated village through terrain full of trees and roots – a rotten, backbreaking job with hand tools. From 9:00 a.m., the sun becomes unbearably hot with no shade, and I hate every minute. Our rationed drinking water is hot, full of slime, inundated with drowned flies, and tastes terrible. I'm sure this is what brought on my dysentery – not a pleasant problem out on this bald desert.

Outward Bound Journal. Feb. 18/62

***

Walking back to our village following our full day in the heat is one of the most trying experiences of my life. I'm sure I will never make it. I'm so exhausted and so thirsty. Never have I come so close to crying while attempting some physical feat. I push the flies and slime aside, fill my water bottle, and collapse onto the mud floor in our hut to get some rest. Immediately, dozens of flies swarm over my body. Sleep is impossible!

Outward Bound Journal. Feb. 20/62

***

In my individual evaluation with the principal, Mr. Snowsell said he's extremely happy. He said he is very happy that I'd been on the winning team in the plateau scheme. He said it is very difficult to convince Africans today that white men are able to, could, and would withstand many hardships. He said that I had certainly proved this point to them from my performance. He said that I work very hard and seem to do very well under the extreme circumstances.

Outward Bound Journal. Feb. 27/62

***

At this fork in the road of my life, I choose to return to Canada and pursue a degree as a Physical Educator, preparing to start Canada's first Outward Bound School.

Outward Bound Journal. Feb. 28/62

I've been thinking an awful lot lately about becoming an instructor in one of these schools. I love the outdoor life so much, and this would certainly place me in a position to follow my love. My aim would be to become the principal of a leadership and citizenship training centre in Canada. I feel the work would be very rewarding since I would be responsible for producing young leaders and better citizens for Canada.

Outward Bound Journal. Feb. 28/62

***

My four weeks at the Nigerian Outward Bound School change my life forever. For the first time, I realise I could be engaged in a profession where I go to work every day engaged in a job I'm passionate about. You see, my dream of dentistry included setting my own hours and making lots of money. I would have enough spare time to embark on those outdoor adventures that are my passion. This was my crossroads. My decision would direct the remainder of my life. Should I pursue my dreams of exploring cavity-filled mouths? Or should I follow a new path and prepare for a fulfilling career which would allow me to follow my passion every day?

Outward Bound Journal. Feb. 28/62

Capstone Activity

Our capstone activity is called the 'plateau scheme'. It's an orienteering exercise through the hills of northern Nigeria. We set our own goals and then try to accomplish them. We set an ambitious goal of hiking eighty-two kilometres, navigating the rough terrain in blistering heat, with heavy backpacks containing all our supplies for three days – encountering the challenges of river crossings, mountain ascents, scree slopes and wild animals. My two Nigerian teammates and myself accomplish our goal and set a new record for the exercise.

Outward Bound Journal. Feb. 26/62

I'm awakened from my sleep as I lie in my sleeping bag on the third day of our capstone challenge, a hand across my mouth.

'Quiet! Not a sound!' my teammate whispers.

There's tension in the air. A primeval scream reverberates through my body. I rise fearfully.

It's like a theatre curtain has been raised. The performance unfolds before my eyes. Black bodies of all sizes. Infants and adults weighing forty kilograms. Hairless faces seemingly carved from old leather. Eyes scouring the savanna for food … and enemies. Menacing fangs as long as seven centimetres. Long, strong arms capable of squeezing the life out of a human.

There they are, a troop of the world's largest apes – about eighty-five baboons passing through not more than twenty metres from my sleeping bag! I'm enthralled by the primates – babies frolicking with one another, adults hugging themselves with long arms, pairs grooming one another.

We had been warned of wild troops of baboons, vicious animals capable of attacking humans. The troop continues to move on as we hide silently behind an outcropping of rock. My time with the baboons over – luckily. I am not relishing an encounter with these big, strong, aggressive primates.

REFLECTIONS. Leaning into the Unknown

At various times in our lives, we face the decision of whether we should lean into the unknown and unpredictable. Should I enter an endeavour when I'm not sure of the outcome? What struggles will I face? Do I possess the skill and psychological capabilities to meet this challenge? Do the outcomes outweigh the risks? How can I gain the courage to go ahead and step in?

The unknown – the first-time venture can be terrifying. At the same time, it can be most rewarding – more risk, more reward. If I encounter failure, what's the lesson? Do I embrace failure as an opportunity to learn – to do it better the next time? At the point of making these decisions, I want to know if anything goes sideways, I'm able to escape without serious physical or psychological injury. If leading, I want to make sure I'll not have to face a parent with the message, 'Your child is not coming home!'

Curiosity is one of the most driving forces in my life. I simply want to taste as much of life as possible. I agree with Aristophanes: 'Man cannot discover new oceans unless he has the courage to lose sight of the shore.' Yes, I could have accepted my hockey scholarship at the University of Michigan and not travelled with my family to Africa. Yes, I could have spent my ten months in Ghana living at home, sitting in my residence, or studying all day in the library. Instead, I accepted the challenge to get out, explore and grow. My parents were such good role models when it came to 'leaning into the unknown'.

I'm so happy I bought a motorcycle, opening the doors to so many African adventures – administering first aid in an isolated village; searching for adventures with Juju priests; volunteering with self-help projects with the Ghanaian government; spending weekends exploring Accra and small villages; hitch-hiking 3,455 kilometres round trip across four countries to attend the Nigerian Outward Bound School. Each one of these African adventures played a part in chiselling who I am today. Each encounter left an indelible mark on my life.

I believe we must not jump into an adventure without knowledge, skill, extensive prep and mental toughness. *Knowledge* is the cognitive awareness of what the undertaking entails – understanding the circumstances and dangers; relying on the wealth of personal experience and trusting partners; tapping into the wisdom of locals; weighing risk versus outcomes. *Skill* refers to our own personal skills – those tools we've developed through practical experience. *Prep* indicates we've done our homework, prepared physiologically and psychologically. *Mental toughness* is our ability to dance with fear – to be able to focus completely on the immediate task, blocking out all extraneous distractions. These attributes can be trained and help, not guarantee, success.

It's important for us to develop street smarts we can utilise ourselves and teach our children. It's important to be flexible and adapt procedures based on the situation we find ourselves in. My decisions are founded on evidence-based research. This is how I make decisions when faced with dangerous scenarios. I want the likelihood of surviving without injury to be most likely. If disaster is predictable, it's often preventable.

Of course, the danger element of not being certain of the outcome is something adding to my excitement. Not all adventures end well. Sometimes, shit just happens, resulting in injury or even death. To date, my decisions have been sound. Yes, I've suffered numerous injuries, but none have altered my lifestyle. Physical and psychological wellness are high priorities when making decisions. I am fortunate to say no one I've been leading or guiding has suffered devastating injuries. It's by painstaking preparation and the grace of God that I am able claim this.

One of my favourite formulas for preparing to deal with mishaps is an exercise I often implement when participating in an activity. I visualise a challenge and think about what I would do … *If a canoe's tipped with two paddlers in this rapid? If a paddler is pinned on that rock?* This rehearsal fills my toolbox with options and prepares me with swift and effective actions.

Sometimes enthusiasm can override common sense and land us in sketchy shit. Even when conditions are sketchy, we may convince ourselves we're good. I live for adventure, but I must understand the potential cost. I must be honest about the hazards and my ego. There is usually fear, escalating to terror that accompanies a mishap. Our composure, discipline, focus and determination are what take us through these storms. We must be extra meticulous in our preparation, looking at every detail with a regimented game plan. We must refuse to think of the past or anticipate the future – we must live in the moment.

I don't see myself as a risk-taker. I assess the risk. If it seems too high to me, I retreat. If the risk seems acceptable, I get to live out my passion. The rewards can be life-changing, but they also can be disastrous. Epic adventures can leave lasting wounds, physical and/or psychological. It's important we have a sane estimate of our abilities to come out unscathed. As Garth Brooks states, 'If you're a soldier, you need a war.' Myself, I'm an adventurer. I need challenges.

Preparation and encouragement are two ingredients leading to success. There are no shortcuts – it takes practice, practice, practice. Physical and mental preparation are key. Cognitively understanding how to execute and then practicing those skills leads to a more efficient response. Learning to play polo, the most difficult sport I've ever played, emphasised these two ingredients. As the manager of the Calgary Polo Club, I was able to go out every day to 'stick and ball' (practice striking the ball while mounted). Having veterans coach and encourage me helped this rookie tremendously. Consistent preparation and encouragement were key factors in enabling me to move on to play internationally. Quitting is not an option on our journey to success. Picking myself up out of the rodeo arena dirt, brushing myself off and learning from my mistakes was the only way to progress. Missing my polo shot did not mean I should no longer play. Instead, reflecting on my errors and correcting them resulted in new execution.

***

Pamela, the Chair of West Island College Board of Directors, happened to be a close friend of mine at the time of the sinking of their 'Class Afloat' 3-masted *SV Concordia* with sixty on board. The ship was about 500 kilometres off Brazil when a microburst (a sudden downdraft) struck the sails, and within seconds the boat went from sailing upright to lying on its side – beginning to sink. All teachers, crew and four dozen students quickly slid into immersion suits and scrambled into life rafts – a drill they had practiced over and over.

After more than a day adrift in the Atlantic, they were rescued by Brazilian Navy and merchant vessels. Intuitively, the students knew what to do – they had run through emergency rescue procedures numerous times. There simply is no shortcut – we must prepare for the possibility of disaster.

A real-life experience like this is never choreographed. As leaders, we must seize these rare occasions as 'teachable moments', realising we may never have this learning opportunity again. As I look back on this scenario, I am so proud these students had the wherewithal to follow their rescue procedures as they managed their fear.

'What am I to say to these parents who are calling me? They know all the students are safe, but they are demanding answers,' Pam panicked.

I explained that those are some of the luckiest youth to have sailed the Atlantic. They were prepared. They had gone through emergency procedures. They reacted in a most mature manner. Each student did their job. Not one life was lost. The result is amazing. None of us could have orchestrated a learning moment more impactful than this. Every one of those parents should be extremely proud of the epic challenge their kids faced and overcame.

'Tell those parents how fortunate they are.'

PRINCIPAL'S REMARKS AND RECOMMENDATIONS

It is not always easy for an expatriate student to settle down in a very mixed community, and where the living conditions are rather basic. It was most commendable that Brad soon integrated himself with the Section, and became a very valuable member of the team. He was very much liked by the other students, and they took an interest in him. In return, he gave most loyal service to the Group, and made a very good effort in all the varied activities of the Course. There is no doubt that he had a very sincere interest in the training, and in its social service potential. He was keen, sometimes too keen, and a little too quick with an idea, which might have worked better with a little more thought. He supported those in the lead, and showed a tactful approach when in authority himself. The Course will have helped him to understand better some of the problems facing a young man in Nigeria, or other parts of Africa, and it will have given him definite experience in community living in one of its best forms. He was just the type of young expatriate that can do much to assist race relationships in a country such as Nigeria.

Date 1. 5. 62

R. E. SNOWSELL,
Principal.

Nigeria Outward Bound

5. Getting Comfortable with The Uncomfortable

'Only those who will risk going too far can possibly find out how far they can go.' **T.S ELIOT**

Mount Baker Releases Her Fury

The call comes mid-afternoon via satellite phone from the ranger. 'Get off the mountain. Avalanches are starting to trigger.'

Although the Mount Baker Ski Resort is closed on this day, we've received permission to head up the mountain to our chalet to work on maintenance. Although I ski the Canadian Rockies regularly, I cannot believe the volume of snow that falls on Mount Baker, holder of the world record for snowfall in a single season. The snowpack is so deep the mountain road cannot be cleared with snowploughs. Huge snow blowers launch the snow up over the snowbanks as they carve their way up the mountain. The result is a channel heading up the mountain with six-metre snowbanks on either side. As you descend the mountain in your vehicle, it feels as if you are bobsledding inside a curvy, icy track.

The Firs is a Christian organisation catering to youth with summer camps and a winter ski programme. I've been hired as a ski instructor, and my wife Margie as a youth counsellor. We live in Bellingham, Washington, this winter, just a short distance from Mount Baker.

Ascending the mountain road in our chained-up our van, we arrive at our chalet. I jump out of the van and sink up to my waist in feather-light powder. The entrance to our chalet is nowhere to be seen. The recent snowfall has buried the first floor of our three-storey chalet. We crawl across the snowpack, climb into a second-story window, pull out our shovels and dig down to the first-floor entrance.

Always on alert for avalanche danger while on the mountain, we hear the call come in. The four of us scramble quickly to gather our personal belongings and skis, lock up the chalet, jump into our van and head down the mountain. My anxiety grows as I look out the window at the walls of snow on either side of our narrow mountain road. I'm not counting on snow forming my casket!

Bammm! Snow engulfs our van, slamming into the passenger side, pouring over the roof and dropping down the driver's side. It sounds like a runaway train overhead!

The van is still upright. We're completely buried. It's total darkness! We sit in stunned silence. As the thundering roar subsides, our van stops shaking violently. I fight to collect my thoughts. *Will rescuers be able to find us before we run out of oxygen? Am I going to die on this mountain? Surely there are more adventurous and exciting ways to die.* I envision my tombstone, 'Brad Kilb. Died of asphyxiation in a Volkswagen van.' It's not the way this adventurer hopes to be remembered.

I carefully survey the windows of the van and discover the driver's window seems to show a hint of light.

'Let's roll down the window carefully and try to claw our way out,' I say.

We roll down the window. I hold my breath. *Will the snow above cascade into our van?* Frantically we take turns scraping away the snow until the sunlight beams in through a small opening.

'Yes, we've done it. We've dug through to the surface of the avalanche,' I cheer.

We keep digging until we can squeeze out through the window onto the surface of the avalanche. It's high fives all the way around.

My fear changes gears as I strap on my skis. Now we must ski down this burial road, walls extending into the sky above, waiting for another avalanche to bury us unprotected! We turn our skis straight down the mountain as we try to schuss down the snow-covered road. Not possible. We're unable to gain much speed due to the low incline and new-fallen snow. I just want to get off this mountain as fast as possible.

Finally, we emerge from the walls of snow at the base of the mountain. My fear of being buried alive has turned to elation – satisfaction that our preparation, past experience and skill base have enabled us to ski another day.

Little did I realise this would be only my first episode with a deadly avalanche.

Caving Terror

'Damn! I'm stuck!' I panic.

My outstretched prone body is wedged between the rock floor, walls and ceiling. I try to lift my head to detect an exit. My helmet is pinned beneath the ceiling. Movement seems impossible. It feels as if the walls are alive and compressing against my torso as I breathe in. *Relax, Brad. Keep breathing.* My fingernails claw across the smooth floor, attempting to bite into the stone. Futile. I frantically dig with my toes. No movement! My panicking brain fights to overcome my claustrophobic terror. *Should I go forward? Should I go back?* My brain cries out, *Why did I agree to come on this expedition? God, if you let me live, I'll never do such a stupid thing again.* I need to pull powerfully, dynamically – leave this coffin with no chance. I fight off the urge to shriek out in horror as I'm wedged into this tiny opening. Then I remember Tom's advice.

'As a last resort, if you find yourself trapped in the passageway, exhale so your chest cavity deflates, allowing you to wiggle through.'

I try to relax and blow out as much air as possible. The earth encrusts my face as I vigorously exhale. Dry dust coats my teeth. My eyes squint, feeling so dry as the powder blankets the inside of my eyelids. I struggle to rub my eyes – no chance. It feels as if my ribs are collapsing. I gulp a short breath in, preparing to exhale once again. My throat fills with chalky, dry grime. I exhale and blow out every ounce of air in my lungs. Yes – it works. My collapsed lungs allow me to painstakingly squirm through, inch by inch, resembling a muffled, grunting and groaning worm. Triumphantly, I emerge into a ten-metre cavern.

It's July 1972. I'm teaching high school Physical Education in Sydney, Australia, and playing hockey. I know you're wondering why I'm playing ice hockey in the middle of summer on an outdoor rink. You must remember that south of the equator, the seasons are reversed. Winter in Sydney has no resemblance to the coldest -20C winter temps in Calgary, with the average Aussie temperature averaging +12C.

Playing ice hockey in Sydney requires some new warm-weather skills. With the warm wind coming in off the ocean across our outdoor rink, the ice has a sheet of water lying on the surface. As a centerman awaiting a pass in front of the opponent's net, my go-to shot is to slap the water-covered ice before I receive the puck, splashing water into the goaltender's face, and then shoot the puck past his blurry eyes. Gooooaaallll!

We're changing in our dressing room following another victory.

'Hey, Ken, did you hear about that new cave they've discovered in the Blue Mountains? How about taking a day to do some underground exploration?' Tom says.

'Count me in,' responds Ken.

'You guys can't leave me out,' demands Colin.

'OK. Next weekend we're heading underground.'

Wow. Caving. I've never tried that. I plead, 'Hey. How about taking this rookie Canuck along? Please.'

'You'll be underground for ten to twelve hours. It's not a stroll in the park. Are you sure you want to face those dangers, Brad?' asks Tom.

'Absolutely. I want my first caving expedition to be with you guys.'

'If you're sure you won't be a hindrance, you can come.'

The decision is either a yes or no – once we drop in and pull our rappel rope, there is no turning back! *Do I go for it, or do I pass? Will I end up being trapped, not able to move, unable to breathe, dying right there!* My brain says, *Don't do it, but my heart cherishes the opportunity to explore our subterranean world.*

Freedom is letting go of bounds and barriers as we hurl ourselves into the adventure of living on the edge – on the mountain, ocean or underground. I want to overcome my weaker self – use extreme experiences to break through the barriers in my own mind. Some of us may get this feeling from going for a walk or taking in the vista of a lake or mountain. In the end, we get paid back. We cannot treat our fear timidly – we must run through the fire to grow. Although I'm terrified, I want to step into the unknown with courage and trusted partners. I've never dropped into a non-commercial cave. Entering the earth's bowels without a guide, relying on the past experience of my friends, on maps and carbide lamps as we discover this mystical world will be a totally new endeavour for me.

Tom, the leader of our expedition, is the captain of our hockey team and one of my linemates. He's my closest friend here in Australia. I so appreciate many of his personal traits – commitment, trustworthy, responsible, always looking out for his buddies and teammates. My trust in Tom has grown immensely over our season. Colin is our goaltender, always reliable and ready to do anything he can to give our team a chance to win. Tim, a defenceman, is one of the hardest workers on our team, ready to give credit to others as he plays his supportive role. I'm psyched to step into the unknown with these three teammates. They're savvy and reliable. They've gained my trust throughout this season on the ice. I believe in them. I believe they will make the right decisions.

***

'Let's go. It's time to explore,' exclaims Tom as we stand at the entrance to our cave, no bigger than a hula hoop.

We secure our 75-metre climbing rope, climb into our harnesses, double-check our gear, light our carbide lamps and drop into the pitch black of the cave entrance. My heart is jumping – I'm not sure from fear or excitement – perhaps a combination of the fear that I might never exit this underground domain and the excitement of scrambling through this new subterranean network. The tar-black darkness envelopes me as I drop into the bowels of the earth. The darkness reels my brain into a new level of fear. I rappel down twenty-five metres, join Tom and wait for our companions on this small ledge just big enough for the four of us. In the absolute stillness of the cave, I can hear the flow of the river below. Humidity closes in around me – it seems to crush my body. As I breathe in through my nose, the scent is musty. The absolute darkness leaves the world beyond our carbide lamps a mystery.

We pull our rope, drive in another anchor as the chamber echoes with our pounding, secure our rope and descend another thirty metres. As I come to rest fifty-five metres below the earth's surface, I feel the water as the underground river soaks through my boots and socks. My feet are drenched. The water's cold. My forehead beaded with sweat. My arms tired from the rappel.

The cave is dead silent – the only sound, the tumbling water. I take time to bring my heart rate down. I look up at the thin shaft of sunlight – the last glimpse I'll get for ten hours. As Tom coils our climbing rope, I realize there's no turning back. We're committed. We've got to follow this underground current until we emerge downstream. I can't help thinking, *What awaits us?*

My carbide lamp reveals the river flowing through cavernous limestone rooms studded with ceiling-to-floor stalagmites and stalactites. I'm in awe of this underground architecture that has taken centuries to form, one drop at a time. As we wade waist-deep down the crystal-clear river, we enter a fifteen-metre grotto filled with colourful, crystallised stalagmites, stalactites, flowstone and towering columns. These stalactites are beautiful limestone formations hanging down from the ceiling, resembling large icicles or draperies. Rising from the floor are the stalagmites, a similar formation reminding me of upside-down icicles. I gaze in wonder at the white flow slabs and the beauty of the speleothems that adorn the cave. It's like walking through an underground castle – every room decorated with astounding formations resembling an abundance of sculptural decorations supported by fifteen-metre flying buttresses.

The river has etched and eroded the surrounding rock, allowing us to follow its course. At times we must turn sideways and squeeze through narrow openings above the water. Other times we get on our hands and knees in the water and duck under overhanging rock. Sometimes we climb through an overhead opening as the water plunges under a wall of rock. These underground passages seem to be guarding the unexplored chambers against intruders like us. The smooth, cool rock feels like silk worn smooth by centuries of running water.

Perched on a rock shelf above the river, we open our lunch and ponder our next move. Our carbide lamps unveil a wall extending from the river to the ceiling of our cavern. The river dives beneath the rock barrier. There is no way over the solid wall. To continue downriver to the cave's exit, we must submerge below the river surface, swim underwater with the current in total blackness and hopefully rise into the next downstream cavern to inhale a breath of air.

Tom checks our maps. He's sure we're in the right spot. The underwater swim should only be about five metres. Tom decides every one of us can master this underwater swim. *How does he know what I can do? He's never seen me swim! He's only seen me play hockey.*

As I stare at the water disappearing beneath the rock wall, I fully understand there are no pockets of air between the river and overhead rock. I must hold my breath until I enter the next downstream cavern. *Does the underwater tunnel actually open up into another cavity in five metres? Can I manage to swim the five metres? Is the tunnel large enough to allow full breaststrokes?* I realise if these challenges are not overcome, this Australian cave will become my tomb – a 20th-century resemblance of King Tut's burial chambers.

Tom takes a big breath, sinks beneath the water, and vanishes into the chute. We listen intensely, hoping to hear his shouts of joy as he surfaces in the next room. Nothing – not a sound. *Did he make it?*

It's my turn! I prepare for my submerged swim, securing my helmet, hyperventilating to enable my deepest breath. My head seems light. My muscles tense. I start to shiver as if I'm in sub-zero Canadian winter. My breathing accelerates. I've committed to the group, but my gut churns with butterflies! *Should I follow through with my pledge to go?*

I'm fully aware of the possible consequences and my freedom of choice. I know the longer I sit and ponder this decision, the harder it will become. It reminds me of my first dive off the high diving board – go now, or I'll never go.

'OK. Here goes. Pray I get to see you on the other side.'

I'm so nervous. I take the time to bring my heart rate down, murmur a short prayer and take a deep breath. As I submerge, my carbide lamp is extinguished. Blackness engulfs me. I see absolutely nothing. I'm not even sure which way is up or down. I'm encased in pure silence. My helmet bumps on the overhead rock. I feel the current pushing. All my muscles are giving 100% – pulling and kicking. Horror swallows my thoughts. The cold water presses on every inch of my body like a hotdog bun squashing a wiener. There's no air to breathe.

The current pushes me farther into the underwater channel. *Is our map accurate? Are we reading it correctly? Will I be able to reach air in the downstream room?* I must swim with all my strength. At this point, I know everything is at stake. No reversing my direction – no turning back. I must dismiss the grip of these dark thoughts

I transition into an inexplicable level of concentration – a level I simply find difficult to produce in any other way. Many of us would become panicked. Instead, I become super alert – like I'm on some drug – with all my senses engaged. I begin to function like a machine. Everything around me fades away. My thoughts focus on my next move. My single-mindedness narrows. I exist only in the present moment.

In my experiences, I address my fear in an ever-escalating progression. Initially, fear fills my body psychologically, alerting me to the dangers of the situation. Second, my fear moves into the stage of psychological terror. It's at this stage that I must act – take action I believe will pull me through based upon my skill set, past experience and mental toughness. I can't give up. Third, I plunge into panic, a stage where I'm now controlled by the beast! My thoughts are not focused, my actions not intentional. Past experience has taught me that I must act while in stage two.

I no longer feel any fear. I'm focused on my precisely executed actions. I must stroke powerfully and dynamically – leave no chance for drowning. My hands pull. My feet kick. My body glides with the current. There is no wiggle room for any slightest slip of concentration. I move in a relentless rhythm, deliberately paying attention to every detail.

I burst through the surface, exhaling victoriously, gasping in fresh air. My body fills with exultation – I've made it.

Finally, after ten hours underground I spot a shimmer of light. As I emerge from our subterranean world, I feel the warm sun beat down. I can't help but be proud. Caving (spelunking) is perhaps one of the most dangerous things I've ever encountered. To venture into uncharted, underground caves is to put your life at risk. There are several uncertainties – will our carbide lamps last; will my climbing skills be adequate; how far is the underwater passage; does it enter a new chamber; will we get lost in this underground labyrinth? There are only a few people who can imagine doing this and even fewer who will dare to try. It is another occasion when

I realise that if I let these opportunities slip through my fingers, I may never get this chance again. Check off another bucket list activity as I thank God for caving with me.

A Day at Our Retail Outlet

As our Alberta Government privatises liquor stores, Bonnie and I purchase an outlet with her brother Darrell and father Ken. The part-time gig allows me to venture into the entrepreneurial retail business. Stocking shelves is done from behind. I take the bottles from their cases, sliding them onto the shelves one by one. I can see through the shelves into the public area of the store, watching our customers as they mill about selecting their bottles. Since I am by no means an expert when it comes to liquor, I'm much happier working behind the scenes.

With so many customers at this busy Christmas time, I keep an eye on suspect clients as they float around the store. My surveillance is rewarded. I notice an individual take a bottle in his right hand and hide it under his winter coat. I watch him as he stands there in a Napoleon-like pose with his hand thrust into his jacket.

It's my first time spotting a thief in action in our store. *What now? What am I to do? Do I approach him and accuse him of stealing? Do I call the police?* I must act quickly.

Having dreamt about being the hero and catching a thief, I'm now not sure I want to be a hero. My entire body is filled with fear. My mouth is dry. My breathing accelerates. My heart beats ever faster. My palms sweat.

I emerge from behind the shelves and walk to the gentlemen – a burly, tough-looking guy of about twenty-five, cap pulled down over his eyes, his long, red hair hanging out. He's dressed shabbily, has powerful shoulders and a scarred face, indicating many battles. He towers over me at about 195 centimetres. My brain fumbles with what to say. Instinctively, I put out my right hand to shake his.

'Thank you for coming to our store to buy your liquor. We appreciate that,' I feebly utter.

The words manage to come out as I attempt to appear confident and sincere. *Now what do I do?* I'm standing face to face, with my right hand outstretched. I await his reaction.

He pulls his right hand out from under this coat to shake mine. The bottle is revealed. He places it in his left hand and accepts my gesture.

'Here, let me take your bottle up to the cashier and you continue shopping. When you're ready to leave, you can pick up your bottle at the cash register.'

Wow, what a feeling of relief as I stroll over to the cashier to drop off his bottle. My step has a bit of a skip in it as I proudly leave his bottle with the cashier. My feelings of fear have changed to those of pride as I feel I've aborted a theft.

I slide behind the shelves to continue stocking as I peer through and watch the thief quietly leave our store without his bottle. I'm so happy I will end my shift and return home to celebrate Christmas unscathed.

***

It's another busy day at our Calgary liquor store. Provincial regulations state that no liquor outlet can sell to an inebriated customer. Occasionally, our cashiers must enforce this regulation as a drunk tries to make a purchase. Our cashiers have a button under the counter they can press if they're in trouble. The button sounds an alarm throughout the store, and every employee in the store must respond.

I'm unaware of the problem unfolding with our cashier as I empty cases and place the bottles on the shelves. The alarm sounds, and my thoughts change immediately. I'm scared! *What's going on? Do I need to go? Why don't I pretend I didn't hear the alarm and keep working here behind the shelves?*

As co-owner of the store, I realise I must protect the cashier – get out from behind these selves and help. Do something.

Emerging onto the store floor, I see my partner Darrell wrestling on the floor as the shocked cashier stands back. Darrell's lying on top of someone, arms and legs flailing as the two of them struggle. *Oh no. This is serious! Perhaps I should've stayed behind the shelves.*

'Brad. Help me. He's got a hand grenade!' Darrell cries.

I jump on top; grab the hand holding the grenade. His arm flails like a rag doll. I can't get a firm grip. I lose my hold. Darrell's got him in a headlock – forcing his mouth away from the grenade.

'Brad, he's trying to pull the pin. Don't let him get the grenade near his mouth!'

'I'll try to grab his arm!'

'Grab that grenade, or we'll all die!'

We continue to grapple. I'm on the floor underneath both. The tang of male sweat seeps into my nostrils. I get a glance of the man's face – the face of a drunk who'd commit murder. He's slipped Darrell's headlock. His arms are free as we fight to gain control. We can't. I can't reach the grenade.

His hand flashes to his mouth. He grabs the pin between his teeth. The pin slides out. He throws the grenade towards the door. It's only seconds until it'll explode! I try to protect my head. I tuck it under my arms … and wait. Now I understand how our brave soldiers feel as they sit in the trenches waiting for grenades to explode – not a comfortable feeling.

Is this worth it? Is fighting with a drunk worth the few cents we'll make on this bottle? Why didn't I buy a business with less danger? Are Darrell and I going to meet our demise under the body of this drunk?

No explosion. No injuries. No deaths.

Suddenly, the drunk starts laughing. I look at him in disbelief as I try to catch my breath.

'It's a dud. You think they sell live ammo?' says the drunk.

The store across the street is a war surplus outlet. The outlet sells all kinds of armed forces paraphernalia, including unarmed hand grenades. Although I've browsed through the outlet many times, I never thought of purchasing a grenade.

As the police escort the drunk out, Darrell disappears into the office. Our cashier resumes her station behind the till. I continue restocking the shelves. I'll bet your job doesn't provide as much excitement as mine.

Schoolyard Fight

'Fight, fight.'

Terrifying words to my ears in this, my first week of teaching physical education at an inner-city high school. I've been assigned to noon-hour supervision today outside in the schoolyard. A crowd of students gathers as I hope I'm not hearing those words. In all my many years of teaching, I've never had to deal with a battle on the school grounds!

Arriving in Sydney, Australia, I had arranged to instruct at the Australian Outward Bound School on the Hawkesbury River. While instructing at the New Zealand Outward Bound School the week before, I received the news that the Australian School has been temporarily closed to allow for a full investigation of the accidental death of seven students. Instead of instructing, I found a job teaching in Sydney.

I could feel it – fear written all over my face as I slowly proceed towards the gathering crowd. My mental response is immediate – *Oh no, why do I have to be the supervisor on this day?*

As I fake a most confident smile, I inwardly try to recall my pedagogy classes – *Brad, look at this as a learning moment.*

My gut feelings are begging me to head for the safety of the staff lunchroom.

My pace dramatically slows as I saunter towards the commotion. My methodology displays my calculated plan – the longer I take to get to the scene of the crime, the more likely it will resolve on its own without any intervention on my part. As the crowd grows, I creep more closely and freeze as I hear the shattering scream.

'He's got a knife!'

My muscles tighten with fear as I wind my way through the tightly packed crowd towards the centre of attraction. The crowd becomes more vocal, and I become more silent. I don't have a clue how I'm going to resolve this. I have no feasible plan. Although I can't yet see the altercation, my only plan is to proceed as slowly as possible, praying the fight will end before I confront the combatants.

As I arrive at the centre of the ring, the students start smiling. There is no fight, no pugilists, no knife. The students have staged the mock battle to test me, the new teacher, to see what my reaction would be. The students cheer. I've met with their approval. I puff out my chest, look my students squarely in the eyes and stand tall with my new-found confidence. Again,

I learn that lessons are often learned accidentally rather than in the classroom. Sometimes when fear strikes, it's better not to rush into something you know nothing about.

World Indoor Beach Volleyball Championships

I'm living a new chapter in my life, a challenging time of coaching elite athletes and teaching at the University of Calgary. It seems to be happening so quickly that my head is spinning. My job description at the University of Calgary is split 50-50. Half of my responsibility is coaching the Men's Varsity Team, the U of C 'Dinos'. The other half involves instructing in the Faculty of Physical Education, specialising in the Outdoor Pursuits Degree Program.

Often, as I jump into a new challenge, I ask myself the question, *Am I qualified to step into this new position?* I try to reason with myself. *Brad, you've developed and instructed a leading outdoor pursuits programme at a private school here in Calgary. You've served as the head coach of the Canadian National Junior Women's Volleyball Team, leading them to a top-seven finish in the world championships. Of course you can do this.*

I've lived life stretching myself with new challenges. I fully believe that unless I'm willing to step into the uncomfortable, I'll not grow. As a result, I search for opportunities that will stretch and challenge me in every aspect of my life. It's a trait I've inherited from my father, a former professional athlete and international expert in technical education. Dad was an exemplary model: 'Grab every opportunity in life. Let nothing slip through your hands.'

As I become more comfortable in my positions at the U of C, I again get the craving to explore new areas. I author a volleyball coaching book that reaches the Canadian status of Canadian Bestseller, *400 Plus Volleyball Drills*. I produce a white-water rescue film, winning the Gold Medal at the Banff International Mountain Film Festival, *Rescues for River Runners*. I become recognised as a white-water rescue expert internationally. I publish a national volleyball magazine with over 19,000 subscribers, *Volleyball Canada*.

***

A fun gig is promoting the successful Molson Brewery Electric Avenue Beach Volleyball Tournament, spreading 100 tonnes of sand on a downtown street and hosting a beach tournament. But I desire to try something a little different in the volleyball world. The athleticism of the Association of Volleyball Professionals excites me. This professional beach volleyball tour focuses on volleyball, but also encompasses the beach lifestyle. I want to become more than a coach. I'd unsuccessfully attempted to launch a Women's North American Pro Volleyball League, but perhaps I could promote the first Indoor World Indoor Beach Volleyball Championships?

Having promoted several outdoor volleyball tournaments, I was ready to take the big leap. I approach my key sponsor, Molson Brewery, to see if they would sponsor the championships. I present my business plan – an indoor beach volleyball tournament highlighting the top eight ranked teams in the world, a muscle-beach contest for male bodybuilders, a women's bikini contest, and a beach party on the sand for all the spectators with a headliner band.

With Molson on board, my next challenge is to find a venue. Barry MacPherson and I approach the Calgary Stampede Board with my idea. I'm ecstatic as the board agrees to lease their arena for the event. Hunter, the board member responsible for leasing Stampede facilities, steps beyond his bounds and secretly agrees to become partners with Molson in sponsoring the event – an unusual offer since the Stampede does not partner in hosting events. They only lease their facilities. With these two challenges overcome, I now look for a supplier who will rent me 100 tonnes of sand and an earth-moving company to bring the sand into the arena and remove it at the end of the event.

As things fall into place, I now face the biggest challenge of all. I must convince the Association of Volleyball Professionals to come to our event in Calgary. Although I've talked with some of the professional players, they are a little sceptical since they've never done an indoor event before. Luckily, I've established a relationship with these players through my international coaching, and they are willing to support my idea. Barry, the Molson rep, and I board a plane for Los Angeles for a meeting with the AVP players and lawyer.

I foolishly asked some of the players where Barry and I should stay while in Los Angeles. As we pull up to the suggested beachfront hotel in Santa Monica, I realise they have a much richer budget than us. We stroll into the lavish lobby and present ourselves at the front desk.

'Oh yes, Mr. Kilb, the AVP has reserved a double room for you. Please fill out this.'

As I look at the contract, I quickly notice the extravagant cost of the room. I stare in shock at Barry and notice his facial expression of disbelief. With little hesitation, I look at the clerk and attempt to summon up my most confident voice.

'I'm sorry. I'm sure we'd requested the suite. This room will not satisfy our needs,' I boldly announce.

Barry stares at me in astonishment. The clerk looks at his computer screen.

'I'm very sorry, but our suite is already booked.'

'Well, I'm afraid we'll have to find another place,' I state.

I turn around, pick up my bag and head out, with Barry following. As we get to the parking lot, we start laughing so hard tears roll down our faces, and we try our hardest not to pee our pants.

'You idiot, Kilb. What if the suite had been available?'

I'd not thought that far ahead but was confident we'd be able to cross that bridge if it arose.

***

On our shoestring budget, we check in to a derelict hotel and make a phone call to Sinjin Smith, a professional player I know well, to confirm that we're in LA. Sinjin confirms our meeting time the next morning and offers to pick us up at our hotel. Barry vehemently shakes his head 'No' – our hotel would be the worst first impression we could give as the promoters of this world championship.

'Thanks, Sinjin, but let's meet at the lawyer's office tomorrow.'

Our cheap hotel room is perhaps the worst hotel room I've ever stayed in during my travels. It's so bad that in the middle of the night, I'm awakened by Barry's shouting, 'Get out, get out!'

I open my eyes to see Barry standing on his bed, shaking his fist, pointing to the door. Some homeless guy had somehow broken into our hotel room. Following his exit, we once again break into hilarious laughter. I fall asleep thankful this is our last night in Los Angeles. We cannot believe the impression we must be giving as the 'high-roller promoters' from Canada.

***

As Barry and I ride the high-speed elevator to the fifty-second floor of this luxury office building in downtown LA, once again we cannot help but giggle. Here we are, two country bumpkins trying to present a first-class impression. Barry is dressed in his suit, me in my beach attire. I am again reminded of my never-ending quest to seek the uncomfortable and grow. We step off the elevator, and I introduce Barry to the players who are waiting outside the lawyer's office. The handshakes and hugs, warm greetings and smiles help boost my confidence.

Am I going to be able to pull this off? Will my friendly players support me? Will the prize money be sufficient for these top players to come to Canada? Is there something that I've forgotten to do? As we sit in the outer chamber, my mind swirls with many 'what ifs'.

My mouth drops open as we enter the opulent office. I notice a huge poster with the words, 'Thanks Leonard, for all your help,' signed Kareem Abdul Jabbar.

'Oh, I also serve as Kareem's agent.'

Wow, I had no idea we're being hosted by one of the most powerful sports businessmen in the world. Leonard Armato extends his hand to shake.

As Barry and I descend from the heavens on the fifty-second floor, we give each other an exuberant high five and begin giggling again.

'We've done it. They're coming,' I exuberantly proclaim.

They agree to everything we propose. I feel so high with this victory I think I could jump out the window on the fifty-second floor and float down to earth. They promise to have the top seven teams come to Calgary for our tournament. We will select a Canadian team for the eighth spot. The prize money will match what the AVP provides for tournaments – $44,000 USD. We feel proud of ourselves for being able to secure the first-ever Indoor World Indoor Beach Volleyball Championships.

MY LOVE AFFAIR WITH FEAR

***

Now the real work begins. As part of our promotion, we stage preliminary competitions in the bodybuilding and bikini contests in local bars under the banner sponsorship of Molson. Our jury is composed of NHL Calgary Flames, CFL Calgary Stampeders, radio hosts, and of course, Barry and myself. For our beach party, we want to bring in the Beach Boys, but they're far too expensive. We settle for second best – the legendary surf-rock duo Jan and Dean. Along with Molson, I secure Chip Wilson and his Westbeach Apparel as our major sponsor. Chip, a former student of mine, went on to establish the internationally successful Lululemon Athletica. The Stampede agrees to open their concessions and provide ushers. I secure a local trucking company to bring in the rented sand, level it and remove it at the end of the event. Both Molson and the Stampede take on marketing. I hire a disc jockey to play appropriate 1960 surf-music hits between points. I'm ecstatic. The pieces are falling into place as I prepare to promote my most audacious event ever. Even the AVP is excited about this first-ever indoor beach tournament.

***

One week before the tournament, the Stampede president and selected board members call an emergency meeting with Molson and myself.

'The ticket sales are not going as expected. You'll lose a whack of money on this event. Our advice is to cancel the tournament,' reports the president.

Hunter, our liaison with the Stampede board, breaks the awkward silence. 'Perhaps I should have told you, but I've committed the Stampede to co-host the event.'

A dead silence falls over the boardroom. The Stampede board members are stunned. The silence continues as the board members look at each other in disbelief.

'Hunter, you know we never co-host events. We simply rent our facilities to promoters,' the president retorts.

'Well, I felt this event is such a winner that I committed to co-host with Molson,' replies Hunter.

I attempt to clarify our situation.

'Unfortunately, Molson and the Stampede will forfeit the prize money of $44,000 USD if we pull the plug. They already have the money in escrow. Molson feels they've spent a lot of money on marketing, and they don't want to pull out. I would advise that we continue.'

The president of the board leans over and whispers in the ear of his colleague. Finally, he looks up. 'It's decided – it's a go! The Stampede board refuses to renege on the commitment Hunter has made.'

As the Stampede board members storm out of the room, Hunter sheepishly utters, 'I sure hope ticket sales pick up.'

***

The day before our tournament, Barry and I are at the venue setting up the courts. The players have checked into their hotel, the concessions are filled with Molson beer, and I get a call from the Molson rep who is picking up Jan and Dean at the airport.

'Kilb, are you aware that Jan is a cripple in a wheelchair? How do you expect him to be the headliner at our rambunctious beach party?'

I had no clue that Jan had been crippled following a violent car crash. It is too late to change plans. Jan and Dean it is.

The fans love it. The volleyball action is the best beach volleyball in the world. The music gets spectators up on their feet, dancing between points. The finals of the bodybuilding and bikini contest bring out the binoculars. The stage is set up to help Jan out of his wheelchair and be standing as the curtains are pulled back for their performance.

Jan and Dean pull it off. The fans stream down onto the sand and dance wildly as Jan and Dean perform. The arena feels like a beach environment with surfboards, windsurfers, boogie boards and fake palm trees scattered throughout the sand. We close the curtains, and Jan collapses back into his wheelchair.

For all of us in that venue, the event is a huge success. As the dust settles, we receive the report on the financial status of the tournament. Molson and the Stampede swallow a combined net loss of $97,000. I'm sad to say that on the Monday following the tournament, Hunter is released from his position at the Stampede. Molson chalks it up to a promotional and marketing expenditure, so Barry retains his position. I search for the reason for our financial disaster. It seems as if we were ahead of our time in introducing indoor beach volleyball to Canada, as well as competing for fans who were more loyal to our NHL Flames team.

***

A couple of years later, I am flying to California and pick up a *USA Today* Newspaper. In the sports section, I see a picture of Sinjin Smith. He's talking about the indoor beach volleyball tournament being hosted in Madison Square Gardens, New York.

'This whole indoor beach volleyball tour began in a place called Calgary, where a crazy guy called Kilb had the courage to promote the first-ever Indoor World Indoor Beach Volleyball Championships.

The idea of a man-made beach is not a new one. There was even an indoor tournament held in Calgary, Canada, a few years ago, but that one turned out to be a bust because of poor timing. It was during an NHL 'Calgary Flames' playoff series, and you know how Canadians are about hockey. All eyes were puckward, and very few spectators showed up.' *Volleyball Mag.Com*, February 26, 2014

Although the promotion was excellent and so much fun, the finances fell drastically short of our goal. Much to the relief of my family, it was the Stampede and Molson whose pocketbooks were hit. It was my first major leap into the world of entrepreneurship, but not my last.

REFLECTIONS. Getting Comfortable with the Uncomfortable

It's interesting, but once I learned to challenge the unknown and assess risk management, getting comfortable with the uncomfortable became much easier – it's a mechanism that can be trained. Anyone can condition themselves to react better under high intensity and stress. Fear has the potential to get me doing the right thing – the ability to slow down time, to perceive high-speed events with an extra bit of relaxation. It's not a natural gift. It's learned. Anxiety may often be mitigated with systematic practice.

Margie's and my decision to circle the globe put us in unfamiliar situations requiring adapting, creativity, suffering, teamwork and exultation. We didn't set out to discover some new place on our globe but to discover the potential we possess as individuals – push the limits and see how far they stretch. No one else is an authority on what our potential is – it's completely within our own control. We only have a tiny amount of time on this planet. What do we want to do with it?

Our voyage was not easy. But it was rewarding as we came out stronger individuals, exploring our physical, mental and emotional limits. Facing challenges never faced before requires a mindset willing to explore and discover – finding solutions requiring 'street smarts'.

A small example is arriving in Sydney, with no place to live for a short time following the closure of the Australian Outward Bound School. We recycled newspapers, hung them over park benches, and slid underneath to spend the night in our makeshift shelter. And to think we were utilising 'tiny house' techniques that early in our lives.

***

Once we take on a challenge not certain of the outcomes, it is often difficult to turn back – descending into the depths of an Australian cave leaves no easy exit. Sometimes the unknown will throw challenges at us, requiring our utmost focus, skill set, adaptability and grit. Often, game plans must change, but it's these encounters that force us to grow. Exploits summon us to utilise solutions we don't know exist – potential beyond belief.

'Don't live life avoiding risk. Learn to manage risk to experience life to the fullest,' I implore my kids.

Learning to cope with the uncomfortable is a lesson that, although fearful, I'm beginning to accomplish. There's no substitute for being curious, of possessing the skill, boldness and courage to enter new adventures – challenging ourselves to grow in all areas of life by stepping into the unknown ready to learn from our successes, downfalls, mistakes and failures. I want to live life with an open mindset – seeking to grow. It's something I wish to pass on to my children and grandchildren through my exemplary actions and encouragement.

Electric Avenue Beach Tournament

World Indoor Beach Volleyball Championships

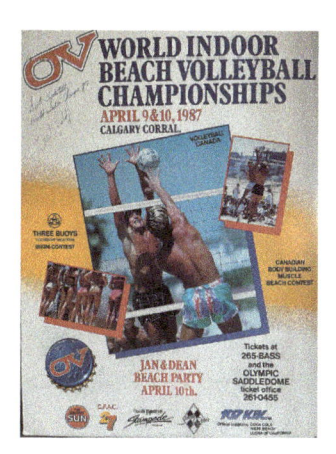

World Indoor Beach Volleyball Programme

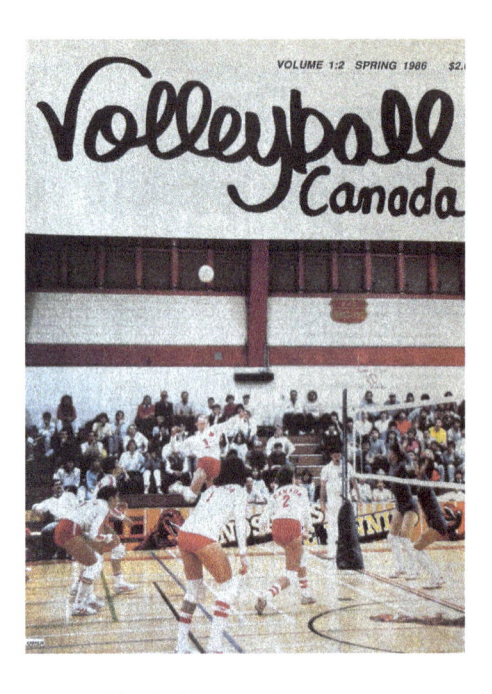

Volleyball Canada Magazine

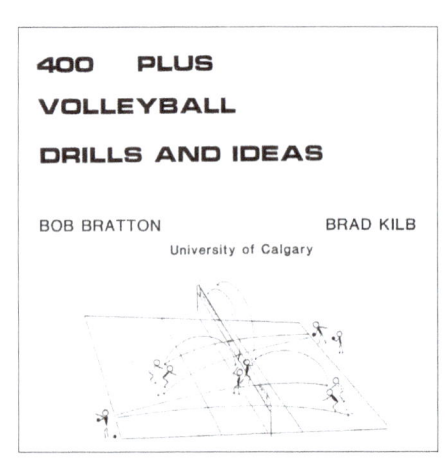

Canadian Best Seller

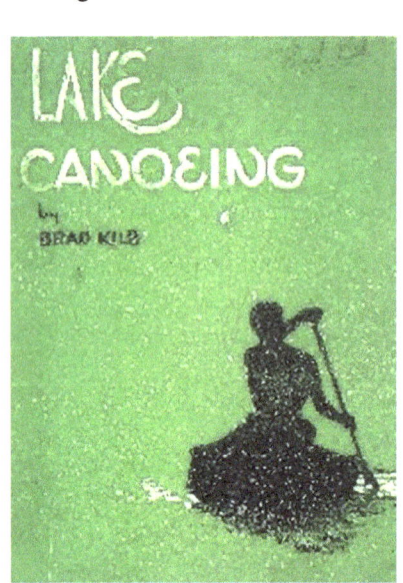

Published Book

6. Creating the Dream

'Excellence is never an accident. It is always the result of high intention, sincere effort, and intelligent execution; it represents the wise choice of many alternatives – choice, not chance, determines your destiny.' **Aristotle**

Surfing to Save My Life

Today my office is in the Canadian Rockies with a group of university students excited about the fresh powder snow as we commence a unit on backcountry alpine skiing. The three-day course is offered in our Outdoor Pursuits Degree Program at the University of Calgary. I am the lucky instructor. The morning newscast has warned of high avalanche danger in the mountains with this fresh dump of snow. My wife Bonnie forewarns me as I walk out the door with my skiing equipment, climbing skins, avalanche transceivers, probes and shovels.

'You've heard the warnings. Whatever you do, *don't* get caught in an avalanche today!'

Avalanches are extremely dangerous natural hazards. The vast, unregulated expanses beyond the patrolled ski resorts are terrain known as backcountry. Many backcountry skiers, snowshoers, snowboarders and snowmobilers lose their lives every year in avalanches.

I gather my twenty students, silhouetted against a cobalt blue sky on the ridge with 275 vertical metres of untouched powder awaiting us. We are surrounded by white, snow-capped mountains thrusting up into the cloudless sky.

I follow the normal pedagogical principles of *experiential* teaching – desired outcomes, safety, explanation, demonstration, activity, and feedback. I can see it in the faces of my students: *Do we really have to go through all these theory sessions? Can't we just start skiing?* We complete an in-depth study of snow profiling – understanding the various layers of snow and the inherent avalanche danger. We then engage in search-and-rescue exercises utilising our radio transceivers – devices that emit signals that can be picked up by other members of the group if an avalanche buries someone.

'Let's ski,' plead my anxious adventurers.

Adrenaline streams through my veins as I understand the enormity of my responsibility – leading these young students into this potential avalanche bowl weighs heavily on me. Skiing in the backcountry is an exercise of knowledge, experience, proper assessment and a leap of faith. It can either reward you with an ecstatic and heart-thumping run or finish you with slow suffocation beneath the heavy snow. *Should I call today off due to the high avalanche warning? Am I sure I'm able to bring all my students home tonight?*

My experience is based on skiing, instructing and guiding over four decades. I've participated in several avalanche courses – learning snow profiling, avalanche predicting, preventative methods to avoid avalanches, strategies for searching for buried victims, procedures to follow if caught in a slide and how to trigger avalanches with explosives and humans. There are dozens of rules only learned through experience. Those experiences stack positive numbers into the equation of life and death. Load the equation with enough positives, and you will have a chance to cross the line of mortal peril.

Still, for all the ability of my past experiences, there will always be a certain element of danger. This remaining risk, the actual threat to life and limb, may be calculated by means of a reasonable, confident estimation of risk factors. How much remaining risk an individual is ultimately prepared to take on is a matter of personal character. Pushing the limits is risky. Risk is, therefore, an integral part of adventure. At times, I am really, really scared, but that is also often when the magic happens. It's the very motivation driving me to seek out situations most people instinctively wish to avoid. It's precisely these extreme ventures that fill my existence with passion and vitality.

Do I allow my fear to overtake me, or do I control that fear? I believe that each one of us is born with an innate capacity to be strong in a crisis, but it can be developed intentionally over time to reach a higher level. There are times for me to act, not panic and give in.

We carve up the untracked, steep, fresh, virgin powder. The kind of snow we all dream about but only see in Red Bull ski movies. Gliding through waist-deep powder, we traverse, carve turns, disappear below the fluffy snow, explode back onto the glistening powder field, gulp a fresh breath of air and submerge once again below the surface, resembling a pod of killer whales surfacing off the British Columbia coast. It's a fantasy-come-true of a fluid top-to-bottom run, soaring down with epic linked tracks. Many say it's like experiencing a piece of paradise – everything is white, and you feel like you're floating. The juices are flowing as we reach the base of the slope. Our joyful voices and hooting reverberate off the surrounding peaks. We're stoked.

Now it's time to go to work as we strap on our skins and start climbing up the slope in the protection of the trees - thick forested slopes offer protection from most slides. As we near the summit, we sit and devour our bag lunches, fired up with the run we've just completed. The chatter is exuberant as we recount the pure elation of floating through bottomless powder. What a way to spend a day at the office, skiing virgin powder in the Rockies.

***

Finishing lunch, we check our transmitters, sling on our backpacks, grab our poles and snap our boots into our bindings. I have a premonition about the changing snow conditions. The sun's beating down on the slope, and it's now 1:30. *Will this snowpack hold? Can I risk another run?* My gut tells me to go out onto the slope solo and make sure the snowpack is solid as the students wait in the protection of the trees.

'Stay in the trees. I want to check the stability of this slope since the sun's been beating down on it,' I state.

I'm sure the students can discern the fear on my face as I take my wrists out of my pole straps. In my years of backcountry skiing, the threatening thought of triggering an avalanche is always present. My confidence wavers as I conjure up negative thoughts. *Should I be exposing these eager students to this danger?* As I step out onto the slope, a student cries out, 'Brad, look above you. The entire bowl has fractured!'

The slope is riddled with a spiderweb of fissures. Tonnes of snow are plummeting towards me. It's too late. There's no escape! I can't get back into the protection of the trees. My heart thumps! My hands turn sweaty despite the cold! My brain races. My breathing accelerates. The hairs on my arms stand erect with goosebumps.

The snow above catapults towards me and suddenly overtakes me. Whoosh – the charging wave of air hits me like an invisible speeding train. *Bammmm!* The full force of the snow slams against my body – it's like getting pulverised by a speeding snowplough. It hurls me forward onto my chest. Miraculously, my head is still barely above the snow, my prone body on the surface of the slide, arms outstretched downslope. It's my worst nightmare – that sense of terror – engulfed in the massive slide careening down the slope. This avalanche is trying to obliterate me as my students witness the snow devour me.

I'm caught in the power of this avalanche. There's no escape – it's like flushing the toilet and watching the contents disappear down the bowl. People talk about the death-defying experience of being attacked by a shark – of feeling the jaws clamping down on a limb. I experience a similar force as this avalanche grabs me with such power that I'm unable to pull away. It feels like there's a vice-like grip tightening on my body.

It's time for me to exercise my ultimate expertise in what I've referred to as 'situational awareness'. My situation is dire – life-threatening. I either panic and allow the slide to dominate me or muster up my cognitive and physical skills to do all in my control to survive. My brain switches into the high concentration zone. I can't die in front of my students!

I frantically throw away my poles. I torque my body sideways, twisting my boots out of my bindings. My limbs are now free. Perhaps I can use them to save my life. Facing downhill, I pick up speed. I've got to keep my head above the snow! I assume the position of a bodysurfer. *Can I save myself by surfing?* With my outstretched arms below me, I can feel the snow churning all around me, rolling under my body and thrusting me up to the surface, picking up speed as I hurtle down the slope, riding the churning snow like a surfer on the front edge of a breaking wave.

As I plunge down the slope on the nose of the avalanche, I become immersed in an otherworldly existence. I become intensely focused on the now – no room for any thought of the past or future. Epic situations are one of those rare moments we become unquestionably immersed in the now. My immediate actions could mean that I live to ski another day.

I'm elated that I'm able to stay afloat as I surf down the slope. The roar of the thundering snow is deafening. My brain is flushed with thoughts, none of which are helping. *Why did I do this? I should have listened to Bonnie.*

Fortunately, the avalanche is not a high-speed slide full of debris. I'm able to surf. It's working. I might live. As I plunge down the slope on the toe of the slide, I spot a tree to my left. *Grab that tree trunk. Maybe I can save my life.* I'm getting closer – get ready – reach out to seize it … missed! My heart sinks as I feel I've lost my last chance to survive. *Don't give up now.* I've got to fight to stay afloat on this slide. I take a deep breath, afraid it may be my last! It's filled with particles of snow. I feel the ice crystals coating the inside of my mouth. I choke.

Nearing the bottom of the slope, the slide starts to slow. I'm buried up to my armpits as I come to a stop, facing downhill. I place my hands behind my neck, enveloping my head with bent arms. If I can form an air pocket, my students may be able to come down the slope, use their transmitters, and dig me out from beneath the snow. I wait anxiously for the snow to engulf my head, burying me completely. *Will I be able to breathe under the snow?* I feel completely powerless.

The snow settles around me, but only to the depth of my shoulders. My head and arms are above the snow. I can't believe it. Most of the snow has chosen another chute.

Until this point, I've not been fearful. I believe it's because I've been so focused on what I must do to survive, reflecting on my training. I'm ecstatic that my head is still above the snow. I'm still breathing – sharp pants of fear. I can feel my heart, constricted by the crushing snow.

Now it hits me! I'm terrified. A blast of panic washes thru my body. I don't want to be entirely buried. I'm terrified my students may ski out onto the avalanche slope and trigger more snow that would bury me.

'Don't ski down the ski slope. Ski down through the trees,' I cry out.

The snow around me is like a concrete coffin. I can't move. I try to take a breath as the snow imprisons my chest feeling like a concrete shell. I'm constricted to gasps. I try to move my legs. Nothing. I can only wiggle my toes inside my boots. I try to move my torso. Nothing. I can hear the beating of my heart in my chest.

As my students gather around me, I can only look up and smile ear to ear with overwhelming relief.

'C'mon, don't stand there staring at me. Please dig me out. Get me out of this tomb.'

I'm utterly reliant upon my students to dig me out! My joy is without restraint – thank you, God. I'm encased in snow rather than a casket. I feel something hit my chest, as if a poke from my brother. I don't care. It must be a student's avalanche shovel, digging furiously. I stretch my arms skyward. They grab me and pull me out. I keep hearing the phrase over and over.

'Kilb, what an unbelievable demonstration.'

Now there may be some lengths to which I'll go for an effective demonstration, but this is not one of them. My instantaneous reactions are a result of my extensive training, and by the grace of God! I truly believe that it is my Creator who has saved me.

***

Please don't get me wrong. An avalanche is not to be tempted. It's a killer that respects no one who gets in its path. Avalanche deaths are not just the stuff of legend; they really do happen. There were thirty-seven fatalities in the USA and twelve in Canada in the 2020–2021 season. Avalanches remind me of a monster in a horror film – they lie in wait for a victim to dare come along. Yes, there are devices that may help a victim remain on the surface of the slide – but there is no guarantee. Yes, there are transceivers that could help to find a buried victim – but victims sometimes cannot be found or could have sucked in their last breath by the time responders find their bodies. Yes, we can take courses on avalanche survival – but so much depends on the type of avalanche, something we have absolutely no control over. The bottom line – if you survive being caught in an avalanche as I was, thank God. You are so extremely blessed to be able to ski another day.

I want to emphasise there were several phenomena resulting in my survival. The slide was a slow-moving slide; was free of debris; travelled down a steady fall line with no drops; most of the snow went down another chute as I came to rest. All these factors were completely out of my control. Even my own rescue was out of my control. I was a victim, needing to be rescued by my students.

***

In the Canadian Rockies, cornices are often formed on ridges at the summit of a mountain. If a dangerous overhanging cornice breaks free, it could initiate an avalanche below. Intentionally triggering a cornice is one method of avalanche control, either with explosives or humans. It's quite a rush to light a stick of dynamite and throw it onto the cornice to initiate the avalanche below.

One of the most exciting thrills I've ever experienced is that of initiating an avalanche using myself as the trigger. Roping up with my partner on belay at the other end of the rope, I hazard out onto a cornice, jumping up and down until the cornice breaks away, triggering a massive slide below. The thundering roar of the snow resonates in my ears. The snow cloud billows into the air. The plunging snow consumes everything in its path, like a humpback whale devouring arctic krill in its cavernous jaws. I pray that my harness will hold me above the deadly slide below. The adrenaline high overwhelms me as I dangle at the extremity of the rope with tonnes of snow thundering down the mountain below me.

During some of these triggered avalanches, we bury a mannequin at the bottom of the slope. As the snow comes rushing down, it buries the mannequin, at which point our instructors would teach and conduct search strategies. Unfortunately, we were not always successful in finding the mannequin. Thank goodness it was only a vanished mannequin – not Brad!

Dog Pile

As a university student, I spend my weekends as a Lake Louise Ski School Instructor. What a gig. I'm following one of my passions as I supplement my scholarships to pay for my tuition. Up Saturday morning; head 180 kilometres west to the Rockies; strap on my boots and skis; and a full day of downhill instructing. Party Saturday night, and start all over on Sunday morning, returning home that evening. As an instructor, I spend the day teaching private and group lessons, but the most fun is a co-ed group of teens I teach both days every weekend. The group is preparing for the race programme, so they are excellent skiers, and my job is to push the envelope with them for an hour and a half each day. We ski the pistes, crush the moguls, shush the green runs, build jumps and head off-piste into the powder. It's a blast.

Powder skiing is my favourite – floating through the bottomless, chest-deep snow – sinking below the fluffy white surface which shrouds my entire body before rising for my next turn – grabbing a breath like a surfacing dolphin before disappearing below the snow once again – hooting and hollering with the kids as we live the powder dream. I'm not sure about the sensation a championship ballroom dancer feels as they glide across the floor, but I'm convinced my exhilaration is close.

'Hey Brad, can we ski this untracked bowl with this fresh dump of powder?' requests one of my students.

'Follow me!'

The great bowl opens at our feet, treeless and white like a giant Roman amphitheatre – wide-open slopes with a luscious coating of white snow like fresh white icing. What a run. We carve up the pristine bowl, shouting and howling like escaping inmates.

The sky is a wintry blue, the wind biting into my face, the snow whipping up into a rooster tail behind, the feeling of freedom. My skis etch a divot into the soft powder as I carve my turns, followed by an explosion as I unweight and spring to the top of the snow blanket, sucking in fresh air before submerging. I'm in a perfect rhythm, in the zone. The world has disappeared. It's only me, the snow, and my skis carving arc after arc.

As I come off a ridge at high speed, breaking through a soft drift, I delight in my airborne position – a moment of suspension – the sensation of soaring as if a bird. I feel a little overbalanced forward and fight to pull back. Arching my back and throwing my head back, I attempt to regain my ideal landing position. No luck!

I hit the snow hanging over my bindings following my six-metre jump. The snow is soft and bottomless, but as I catapult forward spread-eagle, I feel a tremendous blow to my chest. My chest caves in with excruciating pain. I roll over, relieving my chest of my body weight. I attempt to suck in a breath. It's a horrible sensation. I suck in as hard as I can. I can't get air into my lungs. My efforts only result in short, frantic gasps. I've collided full force with a tree stump just below the surface. My chest crushed inwards, my ribs compressing my lungs. My glorious run is cut short as I lie writhing on the snow.

My teens come screaming in, spraying snow over my prostrate body. They gather around, scrutinising me in disbelief. Knowing me as they do, always fooling around with hardly a serious bone in my body, they decide I'm joking.

'Dog pile on Brad.'

The pressure on my chest is so great it stifles my ability to get enough air to speak.

'No, no. I really am injured.' I whisper.

'Yeah sure. Everybody on top of Brad.'

'No, no! Please. I can't breathe! Call the patrol.'

Often my fear is psychological – in my brain. But in this instance, it's due to my excruciating pain and the fact that I can't breathe. The hill medics load me onto their 'blood wagon'.

'It looks like you've probably cracked or broken some ribs,' claims one of the patrollers.

Back in emergency in Calgary, the X-rays confirm the prognosis – two cracked ribs and two more broken, caving my sternum and rib cage inwards. Luckily, I've not punctured my lungs with my fractured ribs. Ten kids lying on top of their instructor would certainly have been fun for my class, but not relieved me of my discomfort!

Swimming with a Humpback

We've been coming to Hawaii with my family for many years. We come in February to align with Reading Week at the university. Luckily, it coincides with the migration of humpback whales to these warm waters for mating and birthing. My fascination with whales is immense, and these trips have afforded many opportunities to observe humpbacks in their natural environment.

On one occasion, Bonnie and I are lucky enough with our boys to see a cow with her calf just offshore as we drive by. She rolls onto her side and brings her fin down with a huge slap, sending water spraying in all directions. It reminded me of the largest cannonball I've ever seen. The baby then tries to replicate the same fin slap. The calf's effort is not nearly as dramatic. His tiny fin drops into the water with barely any splash. Mom demos the action time after time, with the baby attempting to replicate. We watch them together for about twenty minutes. What a sensational experience – mother training baby – maternal instinct working with her young just as my wife Bonnie does.

***

Another time we get a frantic call from a Hawaiian friend. 'Go quickly to a Papawai lookout. A humpback is giving birth.'

We jump into our car and speed off. As we peer down into the clear, turquoise water, we can see the cow in labour. As a mammal, she will give birth to a live infant. We watch as the miracle unfolds before our eyes – baby pops out, surrounded with pools of the life-giving bright, red fluid. Cheers from our small crowd of onlookers. Mom's job done – newborn right at home in his new ocean environment. Mother and newborn relaxing side-by-side in the tranquil waters off Maui.

Then it happens! *No, this can't be true! Please, tell me this is not happening.* But yes, there it is. A large dorsal fin closing in, slowly getting closer and closer. A killer predator stalking its prey. Now close enough to start circling mom and babe.

We know what's below that dorsal fin … a killer shark attracted by the scent of fresh blood. The menacing shark circles and circles, like a merry-go-round at the fairgrounds. The exhausted mother moves her mammoth body as quickly as she can, positioning herself between the shark and her infant – like an impassable wall refusing to let the shark get close to the fraught baby.

The shark is unrelenting. Circling clockwise, then doubling back in the opposite direction. The agility and speed of the predator are far beyond that of the mother – her having just given birth. The predator knows this newborn is about to become prey, as the mother struggles to block his attack. The outcome is inevitable. How can this tired mother continue to be the only defence of her newborn? We watch her movements slowing. Her desire is strong, but her body simply can no longer respond. The jaws of the ferocious shark clamp down on the soft skin of the one-tonne newborn. It's over in a flash! The predator swims away with the victim flailing. The cow humpback lays helpless on the surface of the ocean – exhausted and childless. The laws of the ocean food chain have once again been displayed. It's a fear I am fully aware of each time I dare to enter this alien environment.

***

Although these whale encounters have been incredible, I'm about to experience an adventure even more awesome. On our trips to Hawaii, we've formed a friendship with an expert whale researcher. Every year we meet Richard, a Hawaiian-born environmentalist. Born on his parents' yacht in Lahaina, Richard now spends most of February out in his sea kayak, researching the humpbacks of Maui.

On this trip, we have Darrell's inflatable zodiac and his outboard motor. We head out to find Richard on a calm day, a blue-sky day (a Navy term for endless visibility), the temperature in the high 20sC, a gentle breeze skimming across waves. Three kilometres offshore, we spot Richard leisurely paddling towards us. I cut the motor as I introduce my brother-in-law Darrell to Richard. We spend about twenty minutes updating our lives.

Whoooosh. Silence. Then two more blows. It's the blow of three humpbacks. It's what every visitor to Hawaii craves – an encounter with these majestic mammoths. As they close the gap between us, we realise the whales are accompanied by two dolphins. The curious dolphins come over to our zodiac. They're so inquisitive. We reach out and stroke their heads.

Smooth grey skin. Eyes unmistakably making contact with ours. A smile on their faces. It seems as if they enjoy our human touch as they squeak and click. I can't help but wonder what they're thinking.

Richard explains one of the whales is a cow with a calf, and the third whale is their escort. The escort travels with the pair and hopefully will mate with the female. The trio comes within six metres of our zodiac and silently disappears below the surface. There's no spectacular acrobatics such as breaching, tail slaps, spy hopping or fin slaps. Instead, they stealthily dive out of sight, come up, blow, take another breath and dive once more. We can see their huge bodies underneath our zodiac, gracefully gliding back and forth. Mom with calf in tow, suitor trailing the pair. Richard says the two adult whales are about twelve metres long – the baby about four. These huge mammals entertain us for twenty-five minutes while Richard shoots video from his sea kayak.

As they move off, we excitedly share our encounter with Richard. He's so knowledgeable and fills us with whale stories as Darrell and I listen intently. Suddenly we hear another blow. We survey the waters and spot a fourth humpback swimming towards us. This is a thirteen-metre male pursuing the female who has just left us. He'll try to entice her away from her escort – like the social interaction in any human bar on a weekend. This whale replicates the actions of the previous ones – diving under us and resurfacing. Suddenly he stops, head hanging below his tail – no movement at all. Richard explains this monster is as big as a bus and weighs about forty tonnes.

'Brad, get in the water and swim over to the whale. He's singing to the female, hoping to entice her away from that escort.'

Male humpbacks sing songs to attract a female, trying to convince her that he is THE one. As I take off my PFD, T-shirt and sunglasses, Darrell asks,

'Are you crazy? Don't do it. Do you see what that whale researcher has in his hands? It's a video camera. He just wants a video of a dumb Canuck getting swallowed by a humpback – biblical history repeating itself.'

I've already made up my mind. I'm not going to miss this opportunity. It may never arise again. I roll out of the zodiac and slide quietly into the water. Yes, fear does fill me. My brain flashes with questions. *Is this safe?*

Will this whale turn on me? Will he feel I am interfering with his courtship? Does Richard really know how this bemouth will react? Is Richard just kidding as he instructs me to swim with this whale?

The whale is now only five metres away. The singing resonates even louder as my head sinks below the surface. I can see his immense body. It's even more clear underwater. I stealthily breaststroke to avoid any splash. I get closer … and closer. I see the scars on his back from previous courting battles; his huge pectoral fin, four metres long; the long creases coming back from his mouth under his belly. I move forward towards his eye and look directly into it. I can distinguish his pupil and cornea. He seems to connect visually with me. He looks back at me – his eye scanning my body before returning to meet my gaze. I feel as if I'm part of his ocean environment, as if I belong – no longer a spectator – one of those rare occasions I so seek.

I'm close enough to reach out and touch the hide of this huge male. My hand reaches out hesitantly towards him. *Should I touch him? Would this giant even detect my hand on his hide? How will he react when I poke him? Is he as friendly as the dolphins?*

My fear takes over. I chicken out. I'm afraid to touch even though I'm less than a metre away. I simply hang and soak in the surreal sense of sharing his environment. I continue to tread water beside him until he slowly dives deeper and disappears in chase of his new girlfriend.

I'm ecstatic. I can't thank Richard enough for encouraging me to jump in and explore the unknown; to take the chance; to accept the challenge; to risk my life for this new encounter. I know I would never have gotten into the water without trusting Richard – a trust he had built in our relationship over the years. I knew Richard was an expert when it comes to humpbacks and their behaviour. His knowledge was gained through decades of experiential encounters – not from reading books. Sharing the experiences of his life living with these creatures reassured my confidence in his in-depth knowledge. Trust is an essential component when it comes to overcoming fear.

My Encounter with Sharks

I've completed coaching our University of Calgary women's volleyball team in an international tournament in Israel in 1979. I strongly believe these volleyball tours should include exploration of the country and culture, not merely the inside of hotels and gymnasiums. My wife Margie and I decide to stay after the tournament to see some of the sites of Israel, particularly those surrounding the Red Sea.

We rent a car and head south from Tel Aviv to Eilat on the northern tip of the Gulf of Aqaba. From this point, we travel the deserted desert roads of the Sinai Desert for 280 kilometres, hugging the waters of the Red Sea. Each night we pull up to the shore, park our car and roll our sleeping bags onto the sand – no tent, no shelter. The soft sand provides a mattress, the sound of waves lapping against the shore. The bright moon casts shadows. The night air cool and laced – seeping into our nostrils with a combination of saltwater and desert as we doze off.

Morning arrives as the sun shifts higher into the sky, transforming from pinkish dawn to the hot gold of the desert sun in the azure blue. I lift my water bottle to my parched lips, swallowing gulp after gulp. The cool, life-giving, refreshing fluid slides down my throat, invigorating me for the day ahead. Pushing deeper into the desert between Nuweibaa and Dahab, we see Mount Sinai abruptly rising over 2,000 metres from the wind-swept sands of the desert to our west, while looking to Saudi Arabia in the east. Long camel trains file along the unbroken sands of the desert. Berber tents break the solitude of the dunes. Wooden felucca boats ply the waters for their catch. It's as if we've magically stepped back in history into another world – the world of *Lawrence of Arabia*.

Passing through Sharm El Sheikh, we near the tip of the Sinai Peninsula with the Gulf of Suez to our west. Huge ships glide north and south, reminding us of our voyage through these same waters seventeen years earlier on our circumnavigation of our globe. Loaded cargo ships burdened high with colourful containers. Oil tankers ploughing deep with heavy payloads. The occasional naval vessel, guns visible above deck. Cruise liners appear as drifting Christmas trees against the black night, ghosts hovering above the sand, hulls hidden as they glide silently across the dunes.

There it is – our destination, the underwater Ras Mohammed Nature Preserve. I step onto the beach and survey the landscape – nothing but desolateness. Brown sand to the north, shimmering ocean waters to the south. *How could this place be publicised as one of the most stunning diving spots in the world? Have I been misled? Where are all the divers?*

As we prepare to enter the water, a welcoming park ranger drives up. After the normal greetings, he calmly verifies.

'I'm sure you will encounter grey, blacktip or white tip sharks while diving. If you spot a shark, try not to panic. Hug the sharp wall of coral – try to look as if you are part of the coral. The shark will swim dangerously close to you but will not bother you. They're afraid of a blood-letting injury caused by scraping the coral.'

The look of terror on Margie's face tells me we must quickly drive away from these shark-infested waters. My brain is whirling – full of fear, but at the same time excitement. *Why are we the only ones preparing to dive into these waters? How well does this ranger know sharks? Does he moonlight as a mortician? Perhaps we should head north and look at the magnificent coral and tropical fish on display in an aquarium?*

Are you kidding? I've not driven 1,000 kilometres to pull a U-turn and leave one of the most famous diving sites in the world. However, it's not time to foolishly display my male bravado. It's time to assess the adventure and make a calculated decision. There is a strategy if a shark should appear, according to the ranger. *Will the strategy work? Is the outcome worth the risk? Maybe we should just sign up for one of those shark-cage diving expeditions – at least we would have some protection.*

'Thank you for your advice,' I express gratitude to the ranger.

I've got to make a decision. I fully comprehend the situation. I'm convinced the reward is worth the risk, especially with the ranger's advice. But I have two parties who're frightened – Margie and the shark. I'm able to convince Margie to don her snorkelling gear, but what about the shark?

'Let's go, Margie. We'll never get this chance again. I don't want to miss it.'

The water is crystal clear as we gingerly swim through very shallow water above the rough coral. Then it happens. The bottom of the ocean drops away in a vertical wall of coral, 100 metres deep. I feel as if I'm flying above this world of coral inundated with millions of tropical fish. We've dared to intrude the shark's domain. There's no turning back.

There are so many fish they are colliding with my mask and brushing up against my arms, torso and legs. Schools of so many varieties of fish I've never seen before, not even in my high school biology class. We've entered another world. A world of more than 220 species of coral and more than 1,000 species of fish. My eyes cannot consume it all. My heart explodes with excitement. My brain tries to make sense of this stunning alien environment.

Again and again, I dive down into the midst of schools of fish. We're so engaged in surveying the vertical wall of coral and the astonishing assortment of fish, neither of us has been paying attention to the deeper water outside the reef. Suddenly, I notice a large, dark, image slowly moving back and forth. I have trouble focusing on the image. Is it a shark? The ranger warned us of their presence. Surfacing, I blow out hard through my snorkel and suck in fresh oxygen. I push my snorkel aside and tap Margie.

'Margie. I think I see a shark! We've got to get to the wall.'

Margie's glance of disbelief and terror tells me she should never have followed me into these shark-infested waters. We grasp the sharp coral and pull our bodies as close as we can without lacerating ourselves. The scent of blood would only escalate our shark encounter. The ragged coral fills my body with comfort – the safety of being behind a locked door – out of reach from this menacing predator. Margie's eyes are focused on the coral – she doesn't dare look a shark in the eye. I need to know. I turn to spot what I believe is a shark prowling. *Is it my imagination, or is it a shark?* His back and forth stalking only increases my stress! At one point, I can feel the current wash across my body as his tail propels him past us ever so close. I say a prayer – hoping that Allah understands English. It seems like an eternity. Then as abruptly as he appeared, he disappears offshore.

'Margie, what a mind-blowing experience. Did you see that shark?'

'Are you kidding? I'm terrified. There's no way I'm turning to look.'

I've experienced swimming with humpback whales off the shores of Maui, but this is my first opportunity to swim with a shark. No cage for protection. No support boat to extricate us from the hazardous waters. No guide with a stun gun to ward off the predators. Just the two of us and the safety of a coral wall. Our safety is based on the fact that sharks know enough not to come close to the wall and open up a wound on themselves. The blood would create a cannibalistic shark frenzy, resulting in their death. The thrill of sharing those waters with a shark definitely ranks up there with some of my most adventurous encounters, perceived or not.

I'm indebted to my parents for having instilled within me an attitude of curiosity and adventure, of exploring the unknown, of being willing to venture into the uncomfortable. However, this adventure could have been more uncomfortable than we had planned! Once again, God protected me – that menacing shark decided to snack on some other unfortunate prey. The decision to swim with deadly sharks should not be taken lightly. The wrong decision could end up in a death wish!

Stepping outside of the box is, I believe, the only way we grow – resulting in success or failure. It provides the opportunity for me to understand what I wish to know, not what I already know. It reveals my true and authentic self. It makes me who I most want to be. It keeps me sane in my life. As we all know, words are cheap – actions speak volumes. Adventure is all about discovery – discovery with careful calculations, humble strength, and vulnerability. It's as a choreographed ballet with an amalgamation of preparation, elegance, courage, control and sacrifice – pushing the boundaries of just how far we can explore – incomparable encounters with unpredictable wildlife.

Croc-Infested Waters

'Hey, get your legs up on your board. You'll be taken by a croc!' shouts the dolphin trainer.

I can tell by the look on his face he's not kidding. It's time to get out of this marina.

I've paddled the half-kilometre across the marina. Tired, I've sat down, straddling my board with my legs dangling into the water. Right above me is a tourist attraction where vacationers swim with penned dolphins. One of the trainers has sprinted down the bank to shout his warning.

I've got a half-kilometre stretch of marina to get back to our resort. Let's hope it's not a half-kilometre of hell. Jumping up onto my board, I proceed back. It seems as if my peaceful paddle has suddenly evolved into a stressful nightmare. *Can I get back safely to shore without encountering a voracious croc? Am I able to survive this paddle to live another day with all my body parts and extremities intact?* My eyes widen as I intensely survey the surface for protruding eyes and snout.

I'm on our annual vacation in Nuevo Vallarta, Mexico. Our resort, Paradise Village, is only steps from the ocean, my playground for the month. Swimming, boogie-boarding, boating, whale watching, surfing and standup paddleboarding.

This year I've brought my own inflatable standup paddleboard, so every day I can get out on it. I love the feel of the water under my board; fish jumping alongside; stingrays swimming gracefully under; albatrosses diving for fish; turtles surfacing to take a peek; dolphins playing not twenty metres away; whales blowing and breaching further offshore. The ocean lifestyle is a yearly event I treasure.

I've been playing in the surf for about an hour, enjoying the sensations of this environment – the smell of the saltwater; the warm water lapping over my board and across my feet; cooling off as I fall in; the warm breeze drying off my wet body. Pulling hard, I paddle out beyond the break. Sitting on my board, waiting for the perfect set of waves to ride back in – a new challenge for me. My previous canoe paddling experience helps me manoeuvre my board with similar paddling strokes – sweeps, draws, prys, J-strokes, bow and crossbow cuts. The added stress of balancing on this precarious floating platform is never-ending. My legs are turning to mush as my muscles are incessantly on fire. It's time to get onto some tranquil water.

***

Bordering our resort lies an extensive marina with mammoth ocean yachts, some carrying helicopters. It's another world. The yachting world is something I've tasted only once as I captained a twelve-metre power yacht off the shores of British Columbia. Our marina is completely sheltered, with a flat surface beckoning me to enter. Perfect.

I'm totally aware that crocs infest the marina waters. These reptiles are incredible killing machines with jaws among the strongest in nature. The mango groves lining the river flowing into the marina provide a natural habitat for these predators. Kayaking the river as it empties into the marina, I've encountered numerous crocs sunning themselves on the banks. The adrenaline flow of paddling amid ravenous crocs entices me into the marina.

I can feel my body relax, paddling on this flat surface after fighting the waves for the past hour. Sun baking down on my fatigued body, the peaceful swish of the water against my board as I skim across the surface surrounded by multi-million-dollar yachts. Paddling closely by one of the craft, the deckhand on board calls out.

'It's a good thing you're such a good paddler. This marina is full of crocs.'

I politely nod my head and continue into the marina. *Hmmmm ... I'm not sure I'm that proficient on my board, but I'm not turning back.* As I approach another yacht, I call up to the deckhand polishing the chrome.

'Hey, do you ever see crocs in this marina?'

'Absolutely, but I don't think you need worry. They're only about two metres long, and you should be able to handle them.'

I start thinking as I paddle farther into the marina. *Should I really be doing this?* These guys sound serious. I've seen crocs in the marina. I'm not sure they bother humans, although I've heard of human deaths at the jaws of crocs in this area. *Could I really handle a two-metre croc? Surely it couldn't be that dangerous?*

Now, as I'm paddling back, I discover it! Two eyes breaking the surface a mere fifty metres off my starboard! My mind goes wild. Fear sets in as I fight to manage my rising panic. My breathing becomes faster and heavier. My muscles tense. I start to re-think my decision. Paddling my inflatable, I realise that if the croc chomps down on my board, I will go to the bottom of the lagoon with the deflated board. I quickly bend down, undoing my ankle leash.

But, if I fall off my board, I will have nothing to protect me from those razor-sharp teeth. Bending down again, I secure my ankle leash. My brain is swirling. *Calm down, Brad. Use that self-talk you utilise when coaching your athletes during intense moments. Relax. Breathe deeply. Focus on the fundamentals. Live in the moment. Use your past training to conquer this challenge.*

Planning my escape route, I spot a dock about thirty metres off my port side. *If I can just make it there before the croc reaches me, I'll live!* I seem to be gaining ground on the croc. He's not closing the gap. The dock is getting closer. I'm going to make it. Pulling up alongside the anchorage, I step onto the floating wooden wharf. Safe!

I stand and stare out at the croc, who seems to be lying in wait. It hasn't moved since I first spotted him. As I look more intently, I realise what I thought was a prehistoric killer's eyes above the water is simply a coconut. I begin giggling as I overcome my stress. *C'mon, Brad – terrified of a coconut?*

Reaching our shore, my wife Bonnie and her friends are waiting.

'You idiot. If you go back into those waters, I refuse to visit you in the hospital. I will not be your caregiver after you feed yourself to those crocs.'

Once again, God has looked after me – I've survived.

Controlling Thoroughbred Stallions

'Back! Back!'

I bark out my commands with authority to the excitable thoroughbred stallion in our breeding shed. He is standing tall, rearing up on his hind legs, snorting over me, squealing and striking out with his front hooves. It's the most perilous time of the whole 'teasing' routine – moving the dejected stud back away from the mare. These thoroughbred stallions can be extremely dangerous.

I'm working as the stallion manager at a prominent racehorse establishment in Okotoks, Alberta, upon my return from our world tour. Our ranch is a beehive of activity every day with many different aspects of thoroughbred racing. We're standing five studs, and they are my main responsibility. We are also home to about thirty mares, some of our own and some sent to us for breeding purposes. As well, we have several thoroughbreds we're preparing for the track or are back from the track requiring injury rehab.

Our regulation racetrack is equipped with a three-horse starting gate for training purposes.

My mornings are spent in the breeding shed ... my afternoons much less stressful, breaking two-year-old colts or riding one of our injured steeds down to the Sheep River to stand in the cold, flowing water to help heal his injuries. There're basically two methods of breeding within the equine industry – pasture breeding and hand breeding. When breeding thoroughbred mares, hand breeding is the more common method, avoiding possible injury to the mare and stallion. We only utilise hand breeding.

Handling stallions requires confidence, an understanding of stallion psychology and the ability to anticipate the stallion's behaviour and make quick decisions. Most thoroughbred stallions have an inborn predisposition to attempt to dominate their handlers, other horses, other stallions and mares. Each stallion has an individual temperament and must be handled by an experienced trainer who's able to perceive behaviour which must be corrected before the stallion becomes dangerous. Stallion handlers realise it's risky for the stud to comprehend he's physically stronger and has the upper hand. If such is the case, he may become aggressive, unpredictable and potentially dangerous. The previous stallion manager quit after being seized by his jacket by one of our studs, lifted off his feet, shaken like a doll and luckily flung over the fence. My craving to experience another new adventure triumphs, even though my fear goes through the roof as I hear his story.

I try to build a trusting relationship with each of my studs – always in control, understanding their individual personalities, but working together. The handlers of the mare and the stallion must be coordinated, always aware of the total picture. Teasing and breeding these hot-blooded thoroughbreds every day is extremely nerve-wracking, precarious and dangerous. It's imperative I have control over every stud I'm working with, focusing on my job every minute. Every day as we finish teasing and breeding, I thank God I'm still alive.

I must enter the box stall containing a hot-blooded stud, aware of each horse's temperament. I slide the door closed behind me, preventing the anxious beast from escaping. A loose thoroughbred stallion would be extremely perilous if running unattended on our ranch. He would do battle with another stud until one of them dropped.

Now, it's the 500-kilogram stud and myself alone in the fully enclosed stall with no hasty exit. I must approach the stallion and hook my four-metre lead to his halter. Each one of our five studs has a different disposition. A couple of the studs will turn and face me as I enter the stall, making it very easy to attach my shank to his halter. One of the studs tends to kick and will turn, positioning his head in a corner of the stall, daring me to come up behind him to attach my lead. Another waits until I close the door and then charges, mouth wide open with bared teeth, attempting to bite me in his powerful jaw. With him, I grip my lead with my hands about one metre apart as he charges and thrust the rope into his mouth as he chomps down. With his mouth full of rope rather than me, I now hook the other end onto his halter. I breathe a little more easily as step one is now completed.

These athletes are intelligent. As we start heading to the breeding shed, they know what's awaiting them – an intense date with one of his girlfriends. Entering the breeding shed, the stud is understandably excited, wanting to get to the mare. As a precaution, before I leave the confines of the box stall, I give the command, 'Back.' If the stud does not respond immediately, I strike him with my lead across his shoulder and give him the command once again, 'Back.' I do this repeatedly until the horse is fully focused on me and responds immediately. My fear diminishes as I feel I have control over this horny male.

I lead my stallion to the far end of the breeding shed and back him into his corner. With my stud in position, another handler brings in the mare and positions her behind the barrier. The barrier, three metres long and two metres high, protects the stud from the mare while 'teasing'. Teasing consists of introducing a stud to a mare to see if she is in heat. With the mare standing behind the barrier, I lead my stud forward to the barrier with his head at her tail. A mare who is not in heat will violently strike out with her front feet, kick with her hind feet, and attempt to bite the stallion to keep him away. I must not only pay attention to my hyper suitor, but also be aware of the mare, ensuring that neither I nor the stallion are injured. Although the mare is behind a barrier, if the handler is not paying attention, she can move back from behind the barrier and violently kick with her hind hooves at the stud and myself. It's chaos in this close proximity as two 500-kilogram equine brutes engage in the frantic prelude to mating – snorting, whinnying, neighing, bared teeth and hooves flying.

During the teasing, the vet observes the mare to ensure she is in heat. While his trained eye monitors the mare, I'm ensuring I have control of my anxious, high-strung stallion as he senses his lucky date. Once the teasing is completed, I step in front of my stud and give him the command, 'Back.' Not an easy task as the stud does not want to be separated from his lover.

Often the stud will rear up, towering over me, and strike out with his front hooves – a terrifying position to be in. I must stand in front of the frantic stud, blocking his way to the mare. I must have complete control of my stud, who is much more interested in his companion than me. I'm not sure if he can sense the fear within.

In a flash, the stallion's striking hoof lashes out. I lean back – but not far enough! His hoof catches the brim of my cowboy hat. It tumbles to the ground. My skull still intact, I stand tall as I quiver but control the stud. To this day, I thank God for giving me that extra inch of space that saved my life.

River Rescue

As the coordinator of the Outdoor Pursuit Program at the University of Calgary, the best description of my job is 'going to work and playing all day'. My office changes throughout the year, as I specialise in lake canoeing, white-water canoeing, white-water rafting, alpine skiing, backcountry skiing and wilderness survival. I absolutely love my job as I train adventure guides and leaders in activities I'm passionate about. My focus is skill development, safety, situational awareness, leadership and environmental awareness. Every day in the field provides me with opportunities to influence these young minds as we grapple with the challenges they face. I wake up every morning looking forward to assisting my students in understanding their gifts and the best way to use those talents. The combination of instructing outdoor pursuits students and coaching the varsity volleyball team is a dream come true.

My reputation as an international white-water rescue expert was an outcome I'd never dreamt of achieving. The recognition was very humbling and occurred by accident. It was a process of rescuing victims every weekend as I taught and guided literally dozens of river classes and expeditions across Canada. When paddling rapids in canoes, kayaks, rafts or standup paddleboards, death rarely occurs. Spills, resulting in swimming and a boat rescue, are very common. A paddler may be able to execute very efficient strokes, but incorrect leaning can result in a quick swim. As a result, beginning white-water paddlers may find themselves swimming as they learn. As the lead, I'm responsible for a safe rescue.

As my reputation grew, I became a course director and examiner of National Certification Courses for Professional Canoe and Raft Guides. I would set up rescue scenarios for the candidates to execute. As a result, every week, I was effecting river rescues. This continual exposure to rescue techniques is what gave me the reputation of a rescue expert. I'd never planned on this lofty title.

In 1983, the Canadian Life Saving Society commissioned me to produce a film, *Rescues for River Runners*. The film was an outcome of my white-water leadership and rescue knowledge. It won the Gold Medal at the Banff International Mountain Film Festival. Following the recognition of my film, I was invited to lead many white-water rescue seminars not only in Canada, but also internationally with search-and-rescue personnel. It was an outcome I had never planned to pursue, providing a wonderful growth experience for me as I worked with numerous professional search-and-rescue teams in North America and Europe.

> The number of individual paddlers who would profit by this program would number in the multiple thousands with Rescue Units numbering in the hundreds.
>
> As a professional I have known Brad Kilb both by reputation and in person for nearly fifteen years. His experience in the field of canoe instruction, whitewater rescue, and self-rescue is without peer in the world today. Brad Kilb's breadth of knowledge and his great ability to impart this knowledge to learners is legendary in international whitewater circles. In my travels as vice-president of the National Association for Search And Rescue (NASAR), as a member of the California Governor's Earthquake Task Force, and for more than a decade as the Head of the U.S. Delegation to the International Commission for Alpine Rescue (IKAR), I have been tremendously impressed with the world-wide reputation for instruction of whitewater skills, judgement, rescue, and self-rescue by which this amazing professional is known.

MY LOVE AFFAIR WITH FEAR

***

We're paddling the upper stretches of the Elbow River in Alberta. The course is the Advanced White-water Canoe Course, with every student in the five tandem canoes having completed the Introductory Lake Canoe Course. The students are very proficient with their strokes, but now must learn to deal with moving water. Learning to utilise the fast-flowing water to assist in our manoeuvres, rather than fighting the powerful current, is one of my objectives. Understanding river hazards and being able to avoid them is an ever-present challenge.

The Elbow River is a relatively narrow, winding river exiting the Rocky Mountains as it flows towards Calgary. As I lead my class down the river, I anxiously survey the river as I round every bend. One of the most hazardous dangers on the river is what we call a 'sweeper'. On narrow rivers, the water erodes the bank on the outside of curves washing away the earth below the roots of trees on that shore. As soil erodes away, the tree topples into the river, sometimes lying across the entire width of a narrow river. With the tree lying in the current, the water flows forcefully through the branches and trunk of the fallen tree like a sieve. To understand this danger, think of washing your veggies in a strainer in your kitchen. Any solid object that comes upon the sweeper will be pinned on the upstream side with little opportunity to escape as the current pins the object against the tree. Sweepers are one of the most feared dangers white-water boaters face, particularly during spring flooding.

Rounding a bend in the river, I quickly survey the landscape downstream. To my horror, I see a sweeper across the entire width of the river with no way around. I quickly signal for my students to stop on the left bank about thirty metres upstream from the sweeper. One of the canoes flips, and the two paddlers are dumped into the water. I prepare to execute a rescue as the two students float towards the deadly sweeper. I only have one throw bag – a piece of rescue equipment which allows me to throw a line twenty metres to a swimming victim without getting wet myself. I must not panic.

I realise my predicament. The victims are about three metres apart, eliminating the possibility of a double rescue with one throw. I realise that if I make a throw now, I will only be able to save one of the victims. The other will be swept into the sweeper. *Am I going to watch one die? Which student should I rescue? I've got to find a solution! I'm not taking a student home in a body bag!*

The students are screaming for help. The terror expressed on their faces only adds to the pressure of the situation. I could lose a student if I don't come up with a life-saving solution quickly!

That's it. If I wait as the students are flushed downstream, they will be one behind the other as the triangle flattens out. My stress mounts. I realise if I miss my throw over both victims, I will be faced with a double drowning. *Should I throw now and rescue one student? Can I pull off my risky double rescue – rescuing both with one throw?*

I nervously wait until the moment is perfect. The students are pleading. I make my throw. The bag soars over the top of both victims. They grab the line, both students clutching for dear life. The slack goes out of my rescue line. The line goes taut. I hunker down and get ready for my dynamic belay. The force of both victims threatens to pull me into the current. My belay holds. Both students pendulum to the shore. Saved – three metres above the sweeper.

***

On another trip, I'm not so fortunate. Two of my students capsize above a sweeper, and the current pins them against the tree. I leap out of my canoe and sprint down the shore. I must get to them quickly. I gingerly scramble out on the tree trunk. If I tumble into the current, there will be three victims! I fight to remain calm.

The two students are struggling to hold onto the tree – their chins barely above the water. The force of the current is pulling them downwards under the sweeper. I reach the first victim, grab his personal flotation device, and pull as hard as I can. I've got to get his chest up onto the tree trunk. The water's force is beyond belief – more powerful than I'd imagined. I barely hold his head above the water.

At this moment, the other victim screams. She's farther out on the sweeper. I release my student and scramble down two metres to her. Grabbing her, I pull as hard as I can. Just as I seem to be making some headway, a scream comes from the first victim. Letting go of the second victim I scramble back to the first. *How can I save these students? I cannot give up!*

This procedure continues a couple of times before I decide I'm going to have to push one victim with all my strength underwater beneath the sweeper. If I can push them deep enough, they may be flushed out underneath the sweeper. It's a very unusual, dangerous decision. If I don't push them far enough down, they will be pinned underwater and certainly drown. *Which student do I push underwater?*

Just as I make the decision, I'm able to get one of the victims up onto the sweeper. As he lies exhausted and gasping safely on the sweeper, I'm able to focus on the other. Finally, I pull her up onto the tree trunk. The three of us crawl back along the trunk to the safety of the shore and collapse. I'm completely exhausted, mentally and physically. I say a short prayer, grateful I will return to campus with all my students.

***

On a solid week-long white-water canoe class, we've successfully paddled the upper Red Deer River. The 'Big Red' can best be described as a 'drop – pool' river, with several class-three rapids followed by deep pools allowing for safe rescues of both paddlers and boats. The river is an excellent teaching river from Mountain Aire Lodge with so many sites offering perfect water for learning necessary manoeuvres – upstream and downstream ferry glides, paddling high standing waves, powering through drops, playing in hydraulics, and surfing. We've spent the week paddling and swimming all the challenging rapids – Big Rock, Gooseberry, Jimbo's Staircase, S-Bend, Double Ledge and Cache Hill. I'm proud of the progress my students have made.

As you can imagine, spending dozens of days on rivers leading white-water paddling creates numerous rescue scenarios. It seems that every day I'm on the water, I'm making rescues. Not every rescue is death-defying. Many rescues are straightforward, and I even encourage my paddlers to rescue their colleagues if the rescue manoeuvre is clear-cut. Perhaps the most important lesson I've learned in all these rescues is the importance of prioritising – human lives first, making sure rescuers do not become victims, equipment last. I've also learned that every rescue is different.

Therefore, I developed priorities to follow that could apply to most rescues rather than stringent step-by-step procedures.

***

The Calgary Fire Department has an outstanding water rescue team. They've been on the leading edge of water rescues for many years, particularly rescues involving man-made, low-head dams. Understanding the hydraulics of low-head dams is crucial to making a successful rescue. As the water plummets over the dam, it forms a 'hole' or 'hydraulic' at the base of the drop. A swimmer can be trapped in this 'keeper', forced underwater and pushed away from the dam only to surface in the recirculating surface water pulling them upstream. The recycling water is so strong it will pull a struggling victim or boat upstream to the face of the dam, once again sucking the victim underwater, repeating the cycle until the victim drowns. The only way to escape a hole is to stay down, against all instinct while the oxygen disappears from your burning lungs. You must swim deep below the recycling surface water until you're downstream before resurfacing.

I'm directing an exercise with the Fire Department at the low-head dam in Calgary. We secure a firefighter into a safety harness and let him try to swim out of the recycling water. None of the strong rescuers is able to break clear of the backwash.

***

On another rescue scenario, I am working at the Elbow Falls just west of Calgary. Our objective is to see if we can pendulum behind the waterfalls, tethered on a long line attached to a helicopter. I snap the line into my harness and signal the pilot to lift off. As we gain elevation and I swing below the helicopter, my feeling is similar to that of a Stampede midway ride. I love the sensation of swinging below the helicopter from side to side with nothing below me but air. The most fun is when the helicopter banks with me swinging wildly out from under the chopper.

After a few minutes of joyriding, it's time to get down to work. The pilot positions me directly above the plummeting water of the falls. I signal for him to lower me into the plunging water. The water cascades over my body as I'm lowered below the lip of the waterfall. It's hard to breathe under this torrential volume of water – not nearly as easy as breathing in my shower. As the pilot directs his craft slightly upstream, I swing behind the face of the waterfall. Mission accomplished. With access to a helicopter, we could quickly rescue a victim trapped behind the face of the waterfall.

***

I'm leading a group of university students down the Red Deer River in Alberta. The Red Deer River is an excellent white-water river, with rapids ranging from grade one to three. Grade three is the upper limit of white water for open Canadian canoes. Grade four would swamp the canoe, with the paddlers swimming. The river starts in the Rockies west of Sundre and snakes towards the prairies. The scenery is spectacular, the rapids are excellent, with pools below for safety and rescue, and the wildlife abundant.

'Look, there's a black bear swimming across the river,' an excited paddler claims.

As I look closely, I realise her cub is left behind. As we paddle between the sow and her cub, I'm fearful she may react aggressively since we are forming a barrier between her and her cub. Luckily, she watches us as she rears up on her hind legs but stays put.

***

Coming upon Gooseberry Ledge, we complete a scouting of the rapids and plot our course through the white water. The upper section of the rapid is full of small haystacks. Just below the haystacks, we find ourselves above a one-metre ledge. We must do a downstream ferry glide to river right and drop over the small cascade. The rapid is a tactical run, with little danger. It's a good test for my students.

I'm standing at the bottom of the ledge as the river surges over and fills with air. The aerated water does not afford much buoyancy. One of the canoes plunges below the surface of the river and fills with water. Inside the boat, a student is standing, paddle in hand, water rushing around her legs and screaming, 'Help. Help.'

As I observe the canoe, I realise it's sitting in a shallow spot on the bottom of the river, gunwales barely below the surface. Understanding my student is in no danger, I call out to her.

'The water's shallow. Step out and walk to shore.'

'I can't! I'm too afraid. I can't move!'

Accepting that she's too terrified to move, I walk out to the swamped boat to help. I reach out towards her. She grabs my hand in a vice-like grip, but still cannot move. I'm confused. *Why is she so terrified?* She points inside the boat. A large trout is trapped inside the swamped canoe, unable to escape. The prize trout is swimming frantically inside the boat, brushing against her legs. I giggle to myself as I realise this prize catch has evaded fishers for years, only to be captured in a swamped canoe.

It suddenly dawns on me. The dangers of a raging river are not only real, but sometimes perceived. Perceived dangers are as frightening as those produced by the water.

Speed Kills, but it Sure is Exciting

Speed means exhilaration. My own experience of racing has occurred in four different settings – thoroughbred horses, hydroplane boats, sliding an Olympic track and a Lexus Racing School.

While working as the stallion manager on a thoroughbred racehorse farm, I'm able to 'breeze' some of our racehorses as we prepare them for the track. The feeling of a thoroughbred athlete racing down the straightaway flat out, wind blowing against my face, is so adrenaline-charged, so swift yet so smooth. The muscles of my athlete flex beneath me; I give him his head; my jock saddle so tiny there's nothing to grip; my speed brings tears to my eyes. It's as if I'm racing a Ferrari with the top down.

***

While I was in Australia, one of my hockey teammates is a hydroplane racing fanatic. The pilot of a hydroplane is required to share the cockpit with a co-pilot for safety.

'Brad, how about coming with me this weekend to be my co-pilot?'

'Are you kidding? I'd love to.'

I don my lifejacket, helmet and goggles and climb into the cockpit. I've no idea a boat could go so fast. It feels as if we're flying as we skim across the water, ripping through corners sensing we're about to flip, holding my breath as we plough through waves and rooster tails. It's a lot more electrifying than paddling my canoe.

***

Calgary, the home of the 1988 Winter Olympics, houses a sliding track playing host to the world championships in bobsleigh, luge and skeleton. Bonnie's brother Darrell challenges me to slide the track as his brakeman in a two-man bob. It's too good an opportunity to pass up.

I join the other athletes and warm up in the start shed at the top of the track. Then it's my turn. The Calgary track has a vertical drop of 121 metres with fourteen turns – the most wicked called the 'Kreisel', a 270-degree circular turn with a staggering G-force. Speeds exceed 100 kilometres/hour, and some curves can subject crews to as much as five Gs. With the intensity of the G-forces in the corners, it's often impossible to pull your head back upright.

'Make sure you shrug your shoulders and tuck your helmet down in between them to support your neck,' Darrell instructs.

We push out of the start gate. As we pick up speed, I realize there's no escape from this icy tunnel – no overcoming gravity to slow down; no stepping out of this speeding missile; no stopping. I'm excited, but scared. I falsely believed that as we entered and exited corners, our sled would ride up the icy bank and carve its way smoothly out. Absolutely not true. The sled violently tilts up onto the side of the track and then drops violently back down. No smooth transitions.

As we cross the finish line, I smile and make another checkmark on my bucket list.

***

Our son Brett was invited to a Canadian luge identification camp. At the end of his week, which included a terrible crash, he was invited to join the National Junior Developmental Team. Although he made the decision not to follow through, I decided I must complete my sliding career with runs on a luge and skeleton.

The thrill, sensation and fear of sliding on a small sled is more exhilarating than bobsleighing. I'm so glad Brett talked me into sliding. It's a feeling of speed that puts you so close to the ice, lying on your back or stomach. My experiences of rocketing down the course on a horse, speedboat or car cannot compare. It seems as if you're going much faster than the ninety kilometres/hour. Luge seems a little safer since you're rocketing down the track feet-first. Perhaps breaking your legs is better than breaking your neck?

***

I received an invitation to attend the Lexus Racing School here at the Calgary International Speedway, a 3.2-kilometre track with eleven turns. The invitation emphasised the school's desire to build skill and confidence, with professional drivers working on driving dynamics for performance. As I rocket around the track, my mentor works on the strategies of braking, cornering, maximising the range of my vehicle's agility and a wet pavement drift session. During my in-car instruction, my professional driver refines my skills as I drive the circuit at full speed, talking me through the nuances of racing lines, braking points, throttle application and driving technique. It doesn't take many circuits before I smell the stench of burning brakes! If I tell you I'm not nervous, I'm lying.

'Keep your foot on the accelerator, Brad,' insists my instructor.

At 245 kilometres/hour coming into a left-hand turn at the end of the straightaway, my instinct tells me to brake. Screaming into the corner, terror fills my brain. *I'm going to roll!* Instinctually, my foot eases off the accelerator.

'Don't touch that brake yet,' my mentor reaches over and presses my leg down. My accelerator crushes the floor. We fly around the corner.

'You've got to push the boundaries until we know where the edge is,' he claims.

Late in the afternoon, my instructor exits the car. With angst mounting, my level of focus switches into a different mode – that of a fighter pilot. The course is mine. All my thoughts are intensely centred on driving, nothing else – losing concentration would put me into the wall. My brain recollects what I've learned. At this speed, I've got to respond with instinct.

It's such an exhilarating feeling – it's like having superpowers. I feel so in tune with my car – being one with the car. Until this day, it's been only a dream.

REFLECTIONS. Creating the Dream

'Hi, Brad. How're you doing?' a friend asks.

'I'm livin' the dream.'

'Brad, you're *creating* your dream,' my wife Bonnie chirps.

It's so true. The dream we live is the result of intentional choices – daily possibilities formulating how we live. For our family, lifestyle is of utmost importance – fulfilment earned through experiences, not possessions. I thirst for growth – it nurses my soul. My life is a treasury of personal experiences as I open myself to life, adventure, tragedy and love.

Joseph Campbell said, 'The privilege of a lifetime is being who you are.' As we make choices, we must ask, *Who do we really want to be? Are our choices shaping the masterpiece our Creator created us to be?* We must be willing to peel back the layers of ourselves so we can understand who we *really* are; what our Creator's gifts are; how can we best utilise those gifts; what path we should embark upon. I strive to let this be the compass by which I live my life. Curiosity, ambition, and imagination are enormous gifts that lead me into a life beyond my wildest dreams. 'The biggest adventure you can ever take is to live the life of your dreams.' (Oprah Winfrey)

I truly believe I'm living the life my Creator prepared me for. Some days are painful. I make mistakes. Sometimes I struggle with hearing God's voice as I make decisions. Sometimes I get myself into terrifying shit, physically and psychologically. Sometimes I get jealous when I see the salaries others with my talents are making. I try to be humble as I grow, embracing failure as learning opportunities, transforming me into the masterpiece my Creator created me to be. It's not always easy. In fact, it's damn scary!

Outdoor Guide

International White-water Competitor

Ski Instructor

MY LOVE AFFAIR WITH FEAR

White-water Instructor

White-water Guide

Canoe Instructor

Yacht Captain

7. Terrifying Health Discoveries

'It's your strength and courage that defines you, not your illness.' **Author Unknown**

Battling Malaria

I enjoy my weekly trip to a village close to Accra to administer first aid to the villagers. My basic first aid training came as an outdoor guide – Wilderness and Remote First Aid, Standard First Aid and CPR Certification. I usually spend half an hour playing soccer before playing doctor. The soccer pitch is a dirt lot – not a tree nearby, sun burning our sweating bodies, two sticks upright in the earth for goalposts, shirts and skins to distinguish teams, a ball constructed from woven vines and cheering patients waiting for the goal-scoring 'medical doctor'. High fives with girls and boys, men and women, seniors barely able to keep up. It's such a joyful time.

Administering first aid is my small way of contributing to the community with my minimal skills. As the inhabitants line up, I see everything from scrapes and abrasions to deep blood-covered gashes, infected wounds, fly-infested lacerations, sprained ankles, wonky knees and head wounds.

I'm sure it's the village where I picked up my bout of malaria in the fall of 1961. Although malaria has something of a romantic reputation, let me tell you, it's not at all romantic! The World Health Organization estimates there are 300 million cases of malaria each year, with 1.1 million people losing their lives. When it comes to mortality, malaria reigns as the world's third-leading infectious and parasitic disease.

I feel like an African explorer as I lie hospitalised for a couple of weeks – delirious with high fever, a face shiny with profuse sweating, splitting headache, nausea, vomiting, diarrhoea, muscle pain and convulsions. Malaria is a disease I would not wish on anyone.

Yep, my weekly village visits change me from doctor to patient!

Following Dad's Advice

'Son, I want you to be tested for prostate cancer now that you've passed the age of 50,' cautions my dad.

I know my dad and grandpa both died with prostate cancer – Dad at age 92. I don't mind dying *with* cancer, but I don't want to die *because of* cancer. I decide to go in for that infamous test in early December 1992 – a doctor's finger up my bum followed by a biopsy of tissue! The doc calls me in after he receives the biopsy results.

'Sit down, Brad. I hate to say it, but you've got cancer – prostate cancer,' the doc announces.

I gasp a big breath, blink hard to repress my tears and hesitate a moment to respond with my quavering voice.

'Wow. So, what do we do now?'

'Well, you have a few options. You can ignore the findings. You can submit yourself to radiation. Or you can undergo surgery.'

'Would radiation ensure we get the cancer for good?'

'No, not necessarily.'

'Then let's book the surgery.'

'Surgery is very invasive and could result in some unwanted side effects. I think you should talk this over with your wife before we proceed.'

Dr. Barr explains that with prostate surgery, there are two consequences that could happen, based on the skill of the surgeon – urinary incontinence (the involuntary loss of bladder control resulting in urine leaks) or incompetence impotence (the inability to get or keep an erection firm enough to enjoy sexual intercourse). Neither of these consequences excite me, but I don't need to discuss this with Bonnie. I've already fathered six kids – enough for one volleyball team – I'm not sure I need any more!

'No, I've decided. Book the operating room, please,' I reply.

The doctor calls in his assistant and requests an operating room booking. Before leaving the office, I've got my 'cut and eradicate' date with Dr. Barr. The date of the surgery is January 1993.

I'm not sure how I acquired this attitude that I live by – don't mess around – cut it out before it spreads. Perhaps it's my coaching background? I know it takes just one player to destroy a culture that's been painstakingly built over a number of seasons. So cut that player before they spread their venom!

Driving home full of fear, I make the decision not to share my numbing news with my wife or family until after Christmas – I'm not going to spoil a family holiday with an announcement of something that will be taken care of shortly. I'm so happy Dad insisted I go in for the check-up. It's something I must do with my own boys.

In my preparation for my surgery, I sit down with Bonnie and we write a note to Dr. Barr. I tape it to my stomach before I go into the operating room.

Dr. Barr, please cut carefully.
I have a lot of livin' still to do.
Bonnie and I have a lot of lovin' yet to do.
We're so glad I'm in your skillful hands.

I still see that note framed on Dr. Barr's office wall as I go in for my annual check-ups. He says the entire operating room personnel broke out in laughter as they pulled up my gown to witness the note.

I'm on the operating table for six-and-a-half hours. Dr. Barr reports the surgery's successful, and he caught the cancer before it spread to other organs. I spend a couple of weeks recovering in the hospital before being released.

***

Twenty-four years after I had my prostate removed, Dr. Barr gives me the shocking news that my PSA has risen from 0.0 to an astounding 12.2 – far above the danger threshold. He states that after consulting with other surgeons, he has no explanation. My condition is unheard of. After looking at test results, he states I have two choices – chemotherapy or androgen deprivation therapy. I choose the latter.

After my therapy, my PSA diminishes to 0.0 – great news. I go off the therapy. After being off the therapy for nine months, my PSA is slowly beginning to rise again! I'm confused and frightened. I still have a lot of living and loving left to do. *Why has my cancer returned? Why must I battle with this disease again? Must I live the rest of my life with this lethal disease?* Perhaps once you have cancer, you're never really 'out of the woods?'

When dealing with a disease (or with addiction, I'm sure), sometimes fear raises its head and frightens me to the core. It strikes as I lay in bed, when an unusual pain pops up, when someone drops a 'grief bomb' on me, when I'm reminded of a cancer or heart attack death (like my sons' deaths). It's during those moments that I rely on my faith and reasoning. I will not give in. I will fight this disease with God's help and every ounce of strength I possess.

It's hard to understand the emotions a cancer survivor experiences unless you've been there – fear, anger, sadness – realising you're playing with death. Mindset is so important to keep living. For me, tackling those emotions requires positive thoughts – never giving in to the possibility I might die – only talking about the fact that I'm cured – focusing on the joy of being alive.

Flatlining in the Operating Room

'C'mon, Brad. We're going to be late. I don't want to miss the boys' competition,' Bonnie begs.

'I'm not sure why, but the pain in my chest won't go away!'

'Have you felt this pain before?'

'No. Never.'

'We've got to call the doctor!'

'No, don't bother. It'll pass.'

'How long have you had the pain?'

'Since breakfast – about half an hour.'

'I'm driving you to the hospital, Brad.'

'Are you kidding? We promised to help with the boys' Sport Day at school. I'm not going to miss that.'

In my typical stubborn manner, I don't feel it necessary to rush up to emergency. After much persuasion, I agree I'll go to the hospital for a check-up under one condition.

'I'll only go if I can wear my jogging gear. You drop me off at emergency, and after they clear me, I'll jog the three kilometres to the school.'

As I lay on my gurney in emergency in 1994, the doc shares the news. 'You're suffering a heart attack. The tests show that one of your arteries is eighty percent blocked. You'll be hospitalised until we can book the OR for a simple placement of a stent.'

No discussion would change his mind as I'm wheeled off to the cardiac ward. So much for my jog and Sport Day with my boys!

Days later, I'm wheeled into the OR and watch the stent procedure on the monitor above my operating table. I can see the doctor following the artery in my heart with his instruments. It's quite fascinating. At one point, the doctor announces to the nurse, 'I can't quite see where my instrument is.'

As naïve and trusting as I am, I say nothing. In hindsight, I realise I should've asked the doctor to stop. Unfortunately, he proceeds, breaking through the wall of my artery into my heart. It feels as if someone's plunged the kitchen knife into my heart! I gasp from the excruciating pain. My eyes tear up, my shallow breathing hastens. My fear goes through the roof! *Am I going to die right here on the operating table because of the doc's mistake?* He stops the procedure. No stent. A failed procedure with dire consequences!

That night, lying in the intensive care unit, something goes dreadfully wrong! I break out in a cold sweat. I'm so lightheaded. The walls swirl around my hospital bed. I try to cry out for help. The words won't come. I strain to grasp my help button. My arm won't move. My vision's blurred. I see a nurse struggling with my intravenous injection. My heart monitor signals a change. The signals resemble the sound I've heard on TV – I'm flatlining! The blurred world disappears as I lose consciousness!

As I come out of my blackout, I hear the command, 'Code Blue. Call the Catholic priest. We need him to give Last Rites – immediately.'

It's strange. I'm not sure how long I'd been unconscious – no breathing, no heartbeat. How does a dying person know it's over? I suppose every person is different, but I didn't have an out-of-body event. As I come conscious, my thoughts centre on my survival.

'I can't die. I'm too young. I have six kids and a wife. I refuse to die. I'm going to survive.'

My nurse enquires, 'Can I please phone your wife? We don't think you're going to make it.'

'What time is it?'

'About 2:00 a.m.'

'No. Please, don't wake her up in the middle of the night.'

'Are you sure? I'm a wife, and if I were your wife, I'd want to be here!'

'No, please don't wake her. I'll see her in the morning.'

'Sir, I'm pretty sure you'll not make it 'til the morning!'

I'm exhausted, confused, wondering what just happened. I finally come around, taking deep breaths to make sure I'm still alive, so happy to hear that beep-beep-beep of my heart beating regularly. Those crazy walls remain stationary, and my vision gradually gains clarity – I pinch myself and am so grateful to feel the pain.

My cleared vision permits me to see an individual with a white collar, presumably a priest.

'I'm here. Who needs the Last Rites?'

'Thanks, but I won't need you. It's not my time. I've got too many people depending on me. I'll die another day.'

The priest smiles, makes the sign of the cross and scurries off to welcome another poor soul into heaven.

There are times when we have no answer as to why we are still living. As strange as it may seem, I don't remember fear during this life-altering scenario. Before I flatlined, things happened so fast I didn't have a chance to think. When I became conscious, I was so determined to live fear didn't even cross my mind.

The next morning, Bonnie arrives and reassures me I'm truly still alive. The nurse describes the previous night's events and apologises for not calling while I was experiencing a near-death experience. Boy, when Bonnie hears I'd refused to allow the nurse to call, I sure get an earful! I'm so blessed to be able to squeeze Bonnie's hand. How could such a simple action mean so much? I'm reminded once again that our Creator will decide when it's time.

I'm fortunate collaterals naturally formed around my heart blockage to shunt blood to feed that part of my heart following my heart attack. The medical world discovered how heart disease patients with blocked arteries are able to grow new blood vessels to bypass the blockage. Apparently, not all humans have the ability to build collaterals. Exercise can increase blood flow through these coronary arteries as the inner lining of the arteries responds to exertion by stimulating the vessels to elongate, widen and form new connections. I thank God for equipping my body to do so.

My heart attack occurred at age fifty-three, and I've enjoyed almost three rewarding decades of life ever since. I'm sure each of us has a different explanation of why I'm still alive. Is it due to a wise nurse, my determined and positive attitude or my faith? I believe it's a combination of all. Apparently, we only have four minutes without a heartbeat before we are brain-dead!

I personally believe God also had an omniscient healing hand in sparing me once again and giving me new life with my collaterals. Whether it be guiding that nurse with the appropriate actions, or He Himself laying His healing hand on me, I don't know. However, my faith needs no empirical proof – it is enough for me to believe. I often wonder if it's God speaking to me following my recovery, 'I have much work for you to accomplish, Brad. I'm not letting you die!'

As I look back at my numerous prognoses of personal medical problems, I'm not sure why I've never researched nor worried about the outcome. I put 100% faith in my doctor and my Creator. I've never felt 'it was my time', so I'm able to go into these procedures relaxed and confident. I often feel it's more stressful on my caregiver Bonnie. She's such an angel.

The Silent Killer

I've lived with a small spot on my upper left arm for several months. Finally, in 1995, I decide to have it checked out by my family doctor. Again, I hear that fearsome word.

'You've got cancer – an aggressive melanoma.'

I immediately set up an appointment with my dermatologist. Since the spot is so small, he's able to slice and extricate the skin he believes to be cancerous. The next step is another appointment to confirm the surrounding tissue does not contain any cancerous cells. Again, the bad news is shared with me.

'I didn't get all the cancer. I've got to go wider with my scalpel.'

The cut is deeper and wider, but once again, he's able to sew it up. Yep, you get the drill. The news comes back. He's still not got all the cancerous cells.

I decide to go to Dr. Black (fictitious name), a different dermatologist. The doctor cuts the site bigger and proceeds to take a seven-centimetre slab of skin off my butt. Apparently, there is a shortage of nurses in the hospital today, so the doc engages my wife Bonnie as a fill-in nurse. Although Bonnie is not a nurse by trade, I've watched her sew meticulously in her sewing room. I have more faith in her stitching abilities than she does herself! They work together to stitch my chunk of derriere on my arm. Looking down as they apply the patch to my arm, the thick skin graft looks more like a tire patch than surgery.

I know, you've heard this story before. They send the surrounding skin sample to the lab to ensure that this time we finally got all the cancerous skin. Anxious to hear the results, I can't wait. After not hearing from the doctor for a week, I call his office to update me on the lab findings.

'Hi. It's Brad Kilb, and I'm calling to get the results from my last cancer surgery. Did the doctor get all the cancerous cells?'

'I'm sorry, Brad, but the doctor is away on holidays. He won't be back for two weeks.'

'It's been a week now, and I'm anxious to hear the results.'

'I'm sorry, but the doctor is the only one authorised to give you the results.'

Are you kidding? Three weeks without knowing the results of an aggressive, life-ending melanoma! I'm not willing to wait that long. I have a wife and six children depending upon me. This disease cares not for its victim. Cancer waits for no one. I'm not ready to be another stat as my doctor suns himself on a Hawaiian beach.

I dig aggressively to discover which lab has done my testing and resolutely approach the receptionist.

'Hi. I'm Brad Kilb, and I understand you have the results of my recent melanoma cancer surgery.'

'We may have those results, but I'm sorry, the only person authorised to share those results is your doctor.'

'I understand that, but my doctor is away on holidays for two weeks. While he works on his suntan, my melanoma could be progressing through my body! I must know if he got all the cancer cells.'

'I'm sorry, sir. You'll have to wait until your doctor returns.'

There's no way I'm waiting! I need to know now. I must find out the results. *Do I need more surgery to get rid of this deadly disease?* I barge through the open lab door, paying no heed to the demands of the receptionist. I face the startled technician.

'You're not allowed in here. I'm going to call security.'

Once inside, I turn and lock the door. I pay no attention to the receptionist pounding on the door as I explain my situation to the technician.

'Neither of us is leaving this lab until I receive the results of my surgery,' I threaten.

'I'm sorry, but I'm not authorised to give those results.'

'Whatever. It could be a long wait. It's up to you.'

I try to be nice, but perhaps my resolute demeanour frightens the techy? In a relatively short time, she gives in and shares the results with me. Once again, I learn the doctor has not cleared my body of all the cancer cells! I'm not willing to wait for this dermatologist to finish his holiday with my disease spreading throughout my body.

A friend of mine recommends I visit an outstanding plastic surgeon, Dr. Birdsall. I have no referral, no appointment. I decide to take steps to overcome my fear. I enter his office and am confronted immediately by the receptionist.

'I'm sorry, but I don't believe you have an appointment today.'

I share my dilemma and desire to urgently see the doctor, knowing the cancer is spreading.

'Do you have a referral?'

'No, I'm afraid I don't.'

'I understand your panic, but without a referral or appointment, I can't help you.'

'Thank you. I'll just sit here and wait for the doctor to see me.'

'You're wasting your time. The doctor's not seeing patients today.'

I can tell by the frustrated look and head shaking the receptionist is not impressed. As I sit in the waiting room, two men come through carrying a desk. They set the desk down for a short break. I jump up and ask.

'Can I help you with that desk?'

Both men smile and accept my offer. The three of us struggle but get the heavy desk down the stairs and into the office.

'What are you doing here today?'

Realising this must be Dr. Birdsall, I once again plead my circumstances.

'Come with me.' This empathetic doctor leads me upstairs and addresses his receptionist.

'Book the operating room for me in two days for this patient. Take down all his information while I keep moving.'

The receptionist nods with a dumbfounded look, picks up her telephone and calls. Confirmed. What a relief.

In a very professional manner, with nurses by his side (no Bonnie this time), Dr. Birdsall slices deeply into my upper arm – through all my tissue and muscle. He carves a twelve-centimetre piece of thin skin off my butt and stitches it to my arm.

We wait for the lab results … CLEAN! Success! Thanks, Dr. Birdsall, for stepping in and looking after me. You are my hero.

Wow. I am so grateful I checked out that spot early enough and had the courage to insist those medical boys look after me. It's strange, but I was not fearful as I asked for and prepared for my surgeries. Perhaps it's due to my coaching philosophy – if you have a 'cancer', cut it out before it spreads. As I look at my two options, I'll always choose 'slice and dice'. It's less frightening. It's given me twenty-seven more years of living, and still going.

This experience confirms my belief – there are empathic, compassionate, and excellent doctors out there, but sometimes you need to be assertive and take actions under your own control. Although I've had four more bouts of skin cancer, all my surgeries have been successful. I thank God for our amazing doctors and efficient medical system.

Skateboarding

It's easy to envision. Fantasies flood my brain with actions too vivid to resist. Gracefully gliding from my lecture theatre to the arts parkade on the University of Calgary campus. Nodding politely and waving to admiring students as they observe my athletic ability. I'd love to learn to be proficient at skateboarding.

It's a normal day at work, observing and evaluating students as I lead my kinesiology class in alternative environment activities. We've spent the morning slacklining, and now we've moved on to skateboarding. My class of fifteen students have each donned a helmet, picked up a longboard and followed the student leader to a slight incline. The class has started exceptionally well, with an overview of equipment, safety, and tips on avoiding serious injury when crashing. The tone of our leader's voice seems to indicate that crashing is inevitable. I should pay attention.

I lag behind as the class congregates about twenty metres away. It's my chance. I've got to seize it. Now's my chance to initiate my skateboarding career. I stoop down and pick up one of the surplus boards, not realising this short board is much less stable than a longboard. I stand tall, excited, and drop the board to the pavement – just as I've seen many times on the X Games videos. I'm sure my students will not witness my first attempts as they concentrate on the instructions of their fellow classmate.

I gingerly step onto the board. Place my feet in what seems to be a stable position. My board shoots forward out from underneath me as if shot from a cannon. My backward rotation brings my feet above my shoulders. My head falls back towards the concrete. My reaction is instantaneous – I put my right hand back to meet the concrete. As my hand slams the ground, all my weight transfers through my stiff arm into my right shoulder. An acute wave of nausea wells up from my gut. I gag and swallow back my barf, pant with short staccato breaths and try to slow down my spinning head. As I lie on the ground trying to gather enough courage to stand, I hear a student call out.

'Brad, are you all right?'

I try to focus as a glance towards my class. I barely recognise the fifteen images staring in disbelief. I'm so embarrassed. I can't let them know the extent of my agonising pain.

'Yes, thanks. I'm fine. Just a little tumble.'

I'm relieved to see my students refocus on the task at hand as I attempt to regain my equilibrium and stand. It's not easy. The campus is wildly hurtling around me. Students, trees and buildings a disarray of swirling objects. My spaghetti legs try to support me. My head pounds. My heart beats so loudly I can hear it. I fight back my retching reflex to vomit. I quickly scan my body – no blood. All my trauma lies within my right shoulder, hidden from view as I pretend to be fine.

As the class ends, I try to ignore the excruciating pain in my shoulder as my students offer their sympathy. I walk across campus to the parkade, supporting my limp arm. It's not quite the image I had envisioned before stepping onto that unstable board. Sleep is impossible. There's just no position offering relief from my pulsating shoulder. I follow Bonnie's advice and check in with the doctor the next day.

The diagnosis is a completely severed rotator cuff. No, not separated, but severed.

'Since you're in such good shape, I'd like to wait a month before I decide if surgery is our only option. I want you to undergo intense physiotherapy and check back with me in four weeks,' the doc mandates.

I leave the doctor's office frightened, determined to do everything possible to avoid the knife. I begin physical therapy three times a week combined with over 400 daily reps of rehab exercises. I rigorously work on range of motion, followed by strengthening exercises. It seems to be working.

Although I don't miss a day at work, Associate Dean Gabriele suggests I apply through Workers' Compensation Board for full physiotherapy coverage since my spill occurred during employment hours. I fill out the forms and mail them in. About a week later, I get a phone call.

'Mr. Kilb, I believe you've incorrectly filled in your Workers' Comp claim.'

'That's quite possible. It's the first time in over forty years of employment I've submitted a claim.'

'You've stated your injury was due to a skateboarding accident while working. I'm sorry, but we both know you may skateboard *to* work, but not *while* working.'

'I understand your confusion. I was teaching a skateboarding class as a professor here at the University of Calgary. Teaching this class is part of my job description.'

'You'll have to get a letter from your dean verifying this accident occurred while working.'

A week after submitting my dean's letter, I get another phone call.

'Kilb, it seems as if you've made another mistake on your claim.'

'Oh no. What is it this time?'

'You've made a mistake on your birthdate. No seventy-year-old senior engages in skateboarding. You'll have to send me a copy of your birth certificate.'

I'm sure if I enter the Workers' Compensation offices, I would witness my case posted on the bulletin board under the heading – 'Beware Fake Claims'.

The day of reckoning comes as I enter the surgeon's office. I've done my due diligence. Done everything I could to avoid surgery. Now it's up to him to decide.

'Wow. I'm amazed at your progress. If you promise to keep up with your exercising, I won't need to operate on your shoulder.'

Yes! It's the news I've prayed for. I bounce out of the clinic like an elementary school kid – fist pumping, smiling, hooting and hollering as soon as I get out the door. No surgery. What a relief. Once again, I believe I've experienced Divine intervention.

Three Months of Hell

I've suffered from sciatica for about five years – excruciating pain in my lower back and down both legs. The pain is brutal – like someone hammering a chisel into my back. I'm unable to walk from our university parkade to my lecture theatre without stopping to rest. Standing for my entire lecture is impossible – I must sit on a stool. Enough is enough – it's time for surgery.

Dr. Thomas agrees to go into my spine and expand the opening in C1, C2, C3 and C4 allowing my sciatic nerve to flow freely. December 10, 2019, I go under the knife. I've got such a mix of feelings – excited, optimistic and frightened. One slip of the scalpel and I could be paralysed. An incision eleven centimetres long and eight centimetres deep – through all those layers of muscle I didn't even know I had, right down to my spine. After four hours on the table, I'm rolled out. Shivering. A little groggy. A bit confused. Parched with a dry mouth. A sore throat due to the breathing tube. I can't pee.

'Your vertebrae were a mess – much more than the MRI showed. However, the surgery went extremely well and I'm expecting a full recovery,' Dr. Thomas announces.

A day after my surgery, I'm feeling great – walking around with absolutely no sciatica pain. What a miracle. I'm so excited. My spirits are soaring.

'We're sending you home tomorrow.' Dr. Thomas has done an outstanding job.

I wake up early with thoughts of taking flight home. Not to be. I awake dizzy, head pounding, ready to throw up! I scramble to find my puke bucket, but it's nowhere to be found! I grab the closest receptacle, my bedpan, and puke my guts out. *Oh no, what's gone wrong?* My nurse rushes me down for a CT scan, convinced I have a blood clot in my lungs. The scan comes back – a collapsed lung, but no blood clot. The diagnosis is pneumonia. Hospital release is out of the question. Thirteen more days of hospital food, hooked up to oxygen, a screaming female roommate in constant pain, more blood tests, sleepless nights and visits with Bonnie and son Justin. Not the Christmas I've been dreaming of.

***

One of the perks of my hospital stay is my daily sponge bath. Even at home, I never get to enjoy a back scrub. On this morning, I'm standing butt naked looking out my window as my male nurse sponges my back.

'Hi Professor Kilb.'

I turn around to greet my visitor (turning around was a big mistake). Standing there facing my visitor with full frontal view, I'm shocked to see a former student of mine. I'm speechless – I can't even murmur a 'Hi'. Poised in her nursing outfit stood Jane, wide-eyed with a smile on her face.

'I haven't seen you in years, but when I saw your name on the board, I just had to come by to say "Hi".'

'Wow, I'm not sure but I think it's good to see you again. I'll bet you never envisioned witnessing your professor in this disrobed state when lecturing?'

'You're right, I never did. But perhaps if I had, I would've paid more attention.'

***

It's strange. The sciatic pain down the back of my legs is gone, only to be replaced by excruciating pain in my left leg, hip, and groin. The unexplainable pain's far worse than my previous sciatic pain! Fortunately, my hospital release came in time for me to 'celebrate' (I use that term loosely) Christmas at home with Bonnie. You cannot believe how ecstatic I am to walk into the door of our condo – a great environment, a super caregiver in wife Bonnie and a scrumptious menu. My spirits are lifted once again.

My homestay is short-lived. On New Year's Eve, I'm admitted to emergency to welcome in the New Year in that all-too-familiar elevated hospital bed. After a four-hour wait, I get to see a doctor. X-rays and a CT scan indicate I'm suffering a severe kidney infection. Could anything else go wrong with me? Enough is enough! The pain is brutal – it's like someone plunging a knife into my back.

***

On February 11, 2020, we fly to Nuevo Vallarta, Mexico to join our friends on our annual vacation. I'm anticipating a relaxing, recuperating month – swimming, daily workouts in the gym, biking, stretching and of course sunbathing at our seaside palapa – all activities that will help me rehab. It's two months following my spine surgery and the pain in my left hip, groin and leg continues to be more severe than my sciatica pain.

'Are you sorry you went ahead with your spine surgery?' Bonnie queries.

'The jury's still out on that decision. Give me another month to give the verdict.'

My month in Mexico falls short of my expectations. Although it's a welcome change in scenery from my hospital room, I'm confined to a wheelchair. Yes, I'm able, with Bonnie's help to get to the beach, but my normal activities are certainly curtailed – my daily exercise consists of rolling over in my beach lounge under our palapa!

In the third week of our 'vacation', I'm back in my room violently puking. *Damn! What could this be?* Bonnie insists we go to the hospital. As luck would have it, the chief surgeon is a personal friend of ours. We've played tennis with Dr. Rios and his wife, and enjoyed dinner dates together while in Mexico. What a relief to know he's been trained in the US. He's qualified, I'm confident. He arranges to come in immediately to get to the source of my problem.

The first step is to insert an IV. After two unsuccessful attempts, I notify the nurse, 'I'm sure you know the game of baseball – three strikes and you're out.'

Yep, her third is a strikeout. I ask for another nurse. Can you believe it – it takes five attempts to successfully insert the IV needle. My confidence in the Mexican medical system is painfully waning. Dr. Rios declares I need a CT scan, but the machine is broken. I must travel to the sister hospital in Puerto Vallarta. Bonnie rides shotgun in the front of the ambulance as I'm tossed around on the gurney in the back. The CT scan completed and it's another hair-raising ambulance trip back to Nuevo Vallarta.

Dr. Rios examines the results to discover I have a blocked bowel. I must undergo emergency surgery immediately. Bouncing around in the back of the ambulance, my mind wonders, *Oh no, my worst nightmare – surgery in Mexico!* I'd be lying if I didn't say I was full of fear. I lie there thinking, *Should I go ahead with this emergency surgery?*

Back in the ambulance once again to be greeted by his surgical team, ready to go. Another three hours in the operating room. My surgeon's done a superb job with his scalpel for another successful surgery.

I spend the following five days in the San Javier hospital, a private hospital which is first-class in all aspects. I would not hesitate to be admitted to this excellent facility again if necessary.

On March 9, we fly back to Canada, barely escaping the travel ban put in place due to the Coronavirus pandemic. We move into self-isolation, praying that COVID-19 will not strike. How can I ever thank Bonnie, my caregiver over these twenty-six years of never-ending medical drama – constant crises to overcome and conquer? God has truly endowed Bonnie with unsurpassed strength, love, and patience.

On March 14, I awaken pain free – three months after my spine surgery. It's beyond my understanding. I don't know why I have no pain. I've done nothing differently to generate this miracle. No medical intervention, no new exercises, no pain killers, nothing I can put my finger on. My only explanation is Divine intervention. I truly believe it's supernatural healing – an answer to prayer. God works in gobsmacking ways. He has once again laid His hand on me, and I'm so grateful.

REFLECTIONS. Divine Intervention

As I get older, I realise the most important asset we as humans have is our health. Unfortunately, many of us learn this too late. As a youngster, I lived as if I was invincible. I look back on those early years and get the feeling my Creator must have been protecting me as I foolishly jumped into reckless and perilous adventures. Any one of them could have ended in disaster. When a young adult, I continued to play Russian Roulette, only to be protected by Him again and again. I've concluded that God has a plan for me, and it did not include me passing away at an early age. He protected me, I believe, because He acknowledged I had not yet finished His purpose here on earth. It's comforting, reassuring. It helps me deal with fear in so many situations.

As all of us realise, individual actions are not the only dilemma that can end our life. Along with the not-at-fault fatalities that may occur due to others, we must also overcome countless diseases that may inflict us. I have been inflicted with too many ailments. I remember the numerous times I was diagnosed with cancer – a verdict none of us wish to hear. Waiting for the results was absolutely horrifying. At fifty-two, I am diagnosed with prostate cancer, which returned in my seventy-fifth year. At fifty-three,

I suffer a heart attack. Then just one year after that, I'm diagnosed with aggressive melanoma skin cancer – the cancer, unless caught early, can spell death. Skin cancer has struck me with four more bouts requiring surgery. At seventy-seven, I undergo spine surgery for my sciatica. Then, at seventy-eight, I require surgery in Mexico for a blocked bowel.

Can you believe it – I've survived all. Yes, you could call each one a medical miracle, or is it my Creator who keeps me alive? Not all my family has been so fortunate.

The terror of deathly disease never ends as you rise from the operating table. It continues to plague us throughout our lives – waking up with chest pain, *Is it just a sore muscle or my heart?* Peering into a mirror, *Is that new mole malignant?* Receiving the results from our blood tests, *Does my PSA score indicate cancer?* Each survivor develops coping mechanisms. For me, I have the gift of letting go of the uncontrollable. Don't ask where I got this gift from. I don't know. Absolutely, fear spreads through my body as these triggers appear. But with my constant positive attitude, I'm able to move on without falling into that spiralling downdraft. I believe in the expertise of my surgeons. I know I've done everything within my power to assist my healing. I'm ready to fight through any pain to continue living for my family. I've been blessed to live a life beyond all expectations. I believe God is in control, and I'm ready to accept His decision.

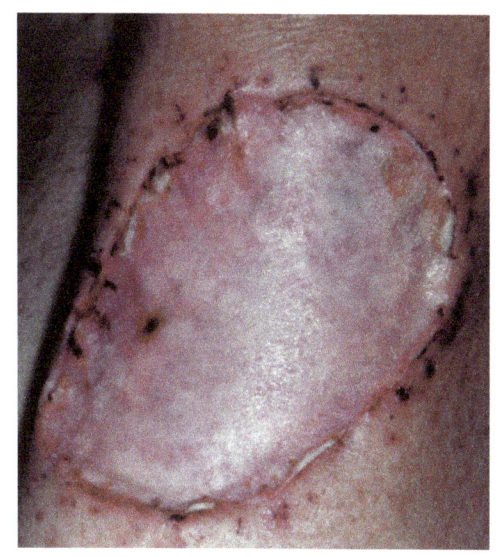

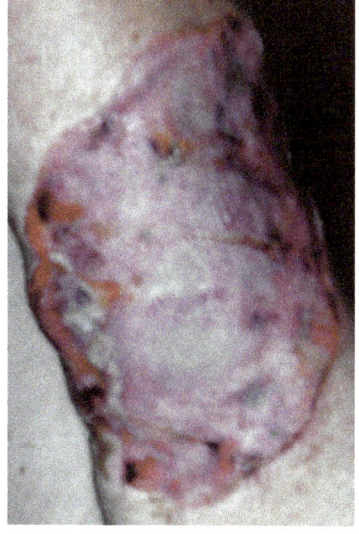

Melanoma, Upper Arm, 1995

Melanoma, 2nd Graft, 1995

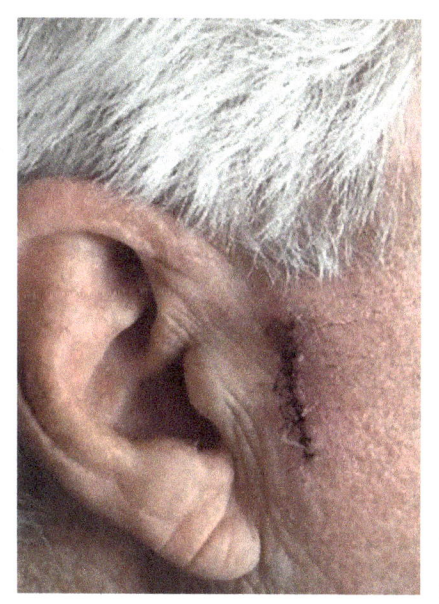

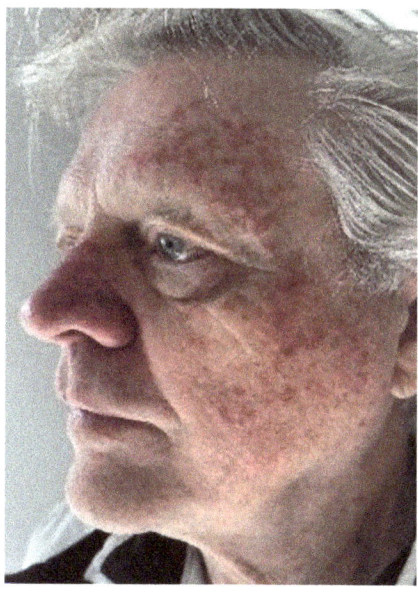

Melanoma, Right Cheek, 2019

Pre-Cancer Face Treatment, 2021

8. Our Creator Will Decide When It's Time

'… the eyes of the Lord are on those who fear him, on those whose hope is in his unfailing love, to deliver them from death and keep them alive …' **Psalm 33:18-19 NIV**

Dealing with Death

Most of us, at some point in our life must deal with another's death. It could be a close friend or family member. The natural order for most of us is to bury our parents, followed by our own death, and finally our children's. In my case, I'm the unfortunate member of the club 'burying our own children'. Organising their 'Celebration of Life' ensues. It's not the way it's supposed to happen, and believe me, it's not easy! I lost two young sons within a period of three years. Although I'd been through the drill once, the second was no easier. However, as you well know, we have little control over when we will leave our planet. That decision is left to a Higher Being – it's often completely out of our control.

I can only speak from my own experience. Each one of us is impacted in a different way and must grieve in our own way. We shouldn't try to influence the way any individual grieves, for grieving is a very personal journey. My wife and I grieve in different ways, without judgement. Her's is not right and mine wrong, nor vice versa. Bonnie helps me daily as I struggle. I've mentioned that I would like to be perceived as a 'man's man' (whatever that means). Throw that fallacy out the window! The fear of breaking down in public is a process which Bonnie has made me more comfortable with. I think my feelings of shame, embarrassment and weakness are the root of this fear. But losing a kid does not dry the reservoir of tears, the anguish that never goes away. To keep those feelings hidden is a mistake – not human. I remember going into my lecture theatre a week after the passing of my son, knowing that I had to tell my eighty-five students of my loss. Yes, my throat dried, my voice quivered, my grief etched on my face and I blinked to hold back the tears, but they flooded as I shared my sorrow. To my amazement, sixty students engulfed me as they hugged, empathised and wept. It was one of the most freeing moments of my life – to be accepted as a human, one and the same, even though I was their prof.

Grieving brings tears at the most unpredictable moments. Certain triggers plunge my heart into deep pain – a song, scent, word or thought. Friends struggle with what to say and say nothing, or innocently drop a 'grief bomb'. Death leaves me with an inner emptiness. The absence left by the loss of my loved ones lingers. There is no moving on, no fix, no cure, no solution for my aching heart, no 'time will heal'. The huge hole remains the same – there is no substitute for our lost ones. The bleeding never stops. It's permanent because my love for them is permanent. For as long as I breathe, I will grieve, love and wish I had more time with them.

However, because I know deep sorrow and continue to claw my way from the depths of unimaginable pain, suffering and loss, I also know unspeakable joy. I daily experience one of the gifts of my faith – that of understanding that I'm able to experience both sorrow *and* joy – it need not be one or the other. Yes, joy as I reflect on the good times we've shared – routine moments that allow lasting memories to flash before my eyes. I now live from a deeper place – with a peace I don't understand. I unconditionally believe my peace is a Divine intervention – a phenomenon that is a gift from God based on my faith. I believe I am stronger today than I was before these deaths. I've had a glimpse into the depths of my soul that would have remained hidden from me without these devastations. This insight has enriched my life and made me more aware of what I consider the purpose of living my life.

Bonnie and I are occasionally questioned, *How could God do this? It's so unfair. Why are you being punished in this way? There are others who deserve to die.* We refuse to go there. We don't have the answers. No matter how extensive the investigation, we will not come up with a suitable explanation. That time spent will only bring more suffering. Can I explain why our faith is healing? No. Perhaps that's why they call it faith.

Every day as I awaken, I have options and choices as I strive to live on without these individuals who mean so much to me. I can live aimlessly without a plan. I can fill my days with thoughts of injustice, unfairness, anguish and anger. Or I can look for ways to feel closer to them. I can be committed to ensuring their spiritual legacy lives on through me. My life can contribute or die with them. It's my choice. I categorically know what they would want.

Lorna – My Mom

Mom was the rock in our family – so loving, so caring, so compassionate, so consistent. Always looking after our needs. I've never seen a couple more in love than my mother and father. What an example for me. As Mom reaches her mid-eighties, her health begins to decline. I fly into London and meet Dad at the airport – no Mom!

'I'm really worried about your mom. She refuses to see the doctor, even though I know her health is failing,' Dad declares as I witness the fear on his face.

Arriving at their apartment, I immediately realise Mom's in serious trouble. We call the ambulance, and Mom's admitted to the University Hospital, diagnosed with major organ failure. Observing a parent's decline in health is fearful. For me, my brain says I'd rather attend their funeral than watch them suffer. It's not a pleasant thought, but I fear even more the thought of a disease ravaging their body and mind.

Wandering into the kitchen the following morning for breakfast, there he is, prostate on the floor. Dad's collapsed and is unconscious! I panic. I struggle with my immediate thoughts. *What do I do? Do I call 911 or follow Dad's wishes?* He's always insisted he didn't wish heroic intervention to save his life. I stare down at his motionless body. *My heart says I've got to do something.* My brain says, *Let him die. This is what he wants.* I'm ashamed I could have these thoughts. *He's my dad. I've got to save him.* I bend down and lean over his face, close enough to feel his warm breath against my cheek. Yes, he's breathing.

What now? Mom's fighting for her life in a hospital room, and Dad's unconscious on the kitchen floor. I dial 911. The paramedics burst into our apartment and cart Dad out on a gurney. The sight of that ambulance driving off turns on the tap – tears pour down my cheeks.

What a disaster. Mom and Dad in the same hospital. With Mom in stable condition, sister Loral and I feel it important not to let Mom know about Dad's hospitalisation. As we're visiting in Mom's room, the announcement comes over the speaker system.

'Would Brad Kilb please report to Albert Kilb's room immediately.'

Yikes! Oh no! *Did Mom hear the announcement? We don't want her to know.* Luckily, she did not catch it.

'Mom, excuse me. I've got to go down to the washroom,' I invent.

I rush to Dad's room only to find they need me to sign a form.

Shortly after Mom's release from hospital, she's diagnosed with cancer. The report is not good. She has a tumour on her brain. The oncologist talks about surgery, but although the result might be life-saving, he feels her quality of life would be drastically diminished with the chemo sessions that would follow. After discussing it with Dad and my sister, we decide not to go with surgery. The risk of an operation would be too dangerous, particularly since Mom's in her eighties. Mom agrees.

I remember sitting in the room with Dad. Mom was sleeping. Dad's weeping.

'Brad, can you come into the hall so we can talk?'

'Of course.'

'It breaks my heart to ask you this, but I must. Can you go down and explain to the nurse that we would like to end your mother's life?'

'Dad. They won't do that.'

'Of course they will. Look at her. We must make this decision to help Lorna.'

'Dad, it's against the law. They can't do that.'

'They treat animals better than this. I can't stand watching her like this.'

'OK. I'll go down and ask.'

Of course, Dad's wishes are not considered. As I walk back to Mom's room, do I fear telling Dad that his request could not be considered? No, that's not the fear consuming my body. It's the fear of finally realising that Mom's life is about to end. Of course, I've thought about losing Mom and Dad, but to see it happening before my eyes … wow! It's a fear that my heart finds difficult to cope with.

Mom's disease advances much quicker than any of us had anticipated. She is bedridden, drugged and therefore not suffering. She spends most of her day sleeping. Mom passes away within the week. I'll never forget the anguish my father went through as he watched Mom slowly slide to her death. To say that Mom was Dad's crux for living would be an understatement.

Albert – My Dad

Dad's relationship with Mom was so deep and loving that he didn't wish to live a day after she passed. The joy left his life. His love of six decades had disappeared and from that point on, he simply existed, not lived.

'Dad broke his hip three days ago, but the seniors home refused to hospitalise him until this morning. I think you should come down, Brad. Dad only has a few days left,' Dad's incredible caregiver, my sister Loral, blurted out the message.

The next day I fly east. The fearful 'beast' is death itself. I must be ready to face it once again. The death of a loved one impacts each one of us differently. I remember going to my best friend's funeral and viewing his body in his coffin. It was such a shock! He didn't even look like the person I knew – his face so distorted, his colouring so white! To this day, I'll not gaze at a body lying in rest. I want to visualise an image of that person at the peak of their life. Perhaps that is why I didn't witness the bodies of Mom, Dad, Brett or Brad Jr before they were cremated.

Little do I know it is to be one of the most heart-rending, but richest weeks of my life. Dad is so sharp mentally, and we reminisce about adventures we had shared. He remembers the smallest details as we laugh and cry together – his body failing, but his spirits high. He's so engaged in every story. I relish that time as I watch his face brighten – something I had not witnessed in the years since Mom passed. Our communication is on such a rich and deep level. The joyful storytelling continues for two days.

On the third day, I see a devastating decline in Dad's health. He finds it difficult to speak, hard to focus, problematic to breathe. I feel the end is near. As hard as it is to share with you, I feel relieved that Dad's time is up. He's not living the life he wants to live without Mom. He wants to die, and I share his feelings. I too hope he will pass quickly without any further suffering.

On the fourth and fifth days, Dad's in bad shape. Our only communication is through touch – responding with a squeeze of his hand. His voice is gone – he no longer speaks. His breathing stops for thirty seconds, followed by a deep gasp for oxygen. It's tough to watch. I ask to see a palliative care nurse.

'Do you have any idea what your father's wishes are?' she asks.

'Yes. Dad undeniably wants to pass away. There's no question.'

'I want you to ask your father what *his* desires are.'

'Dad, can you let me know if you want me to take you off life support? That would result in you passing away more quickly?'

There was no verbal response. He decisively and resolutely raises one arm with his thumb up.

'Wow. That was a pretty clear sign Brad,' the surprised nurse responds.

'Dad, I'm going to ask you again. I want to make sure you answer with what *you* want, not what you think I want. Do you want me to take you off life support? It means you will die.'

Again, no verbal response. This time Dad raises both arms into the air with two thumbs up.

'Brad, you are so fortunate to get such a distinct signal. Have you discussed this with your siblings?'

'My sister is right here, and I know she agrees. My brother would have no say since he severed relationships with my parents decades ago.'

'I believe you should request the hospital take him off life support,' she concludes.

'I'll do that immediately.'

I walk out of the room, shaking like a leaf in a windstorm. I don't want to be the person who ends Dad's life, but I'm aware of his desires. His irrevocable and clear indication gives me no choice but to give the order. My brain tells me it's what I should do, but my heart hesitates. Sometimes cognitive reasoning and emotions do not coincide. I give Dad's request to the head nurse.

Loral and I go to the cafeteria, fumble through lunch together, tears rolling down our cheeks.

'Loral, what did I just do?'

'As hard as it seems, you did the right thing, Brad. It's what Dad wants. Do you think we should have consulted Brian?'

'No. He severed family relationships shortly after his second marriage. Remember how cruel he was to Mom and Dad? His accusations and verbal abuse would send Mom to bed weeping. His wife convinced him that he didn't want to be a Kilb. I'm honouring his wishes – he has no say in what we decide is best for Dad.'

As we walk into Dad's unit following lunch, we run into the doctor who's been caring for Dad.

'Are you the son who gave the order to starve your father to death?'

Wow. It's like a primeval warrior has pierced my heart with his spear. I stand horrified, shattered, dumbfounded, unable to utter a response. I carry on in a fog to Dad's room, the shock of the doctor's statement lingering. My brain seems to abandon my body. *What have you done? How could I end my dad's life? Am I sure it's the right decision?*

Two days later, Dad passes away on his ninety-second birthday. I'm so grateful Loral called, enabling me to spend those days with Dad. I'll cherish them and him forever.

John Ross MacRae – My Nephew

John Ross Macrae was one of the top motorcycle racers in North America. He came by his passion naturally – his father, Doug, owns Blackfoot Motosports in Calgary, one of the top outlets in the nation. JR was like a son to Bonnie and I. He grew up with our boys. Same age. Same interests. Same passions. All three competed locally, provincially, and nationally – JR in Superbike, Brett and Justin in Motocross.

In 2013, JR was racing in the final round of the Superbike Championships in Mount Tremblant, Quebec. He was hungry to continue his rise to the top after winning the previous race two weeks earlier. He came into corner one, overtaking three bikes, losing control, and crashing. At age twenty-four, his racing career and life ended. What an unimaginable blow to our extended family. I never fully understood my brother-in-law's grief until it was my own.

Brett – My Son

It's my first day lecturing in the 2016 fall term. Before I leave for campus, Bonnie and I Facetime Brett and his wife Jessica. We hang up full of delight and excitement as they finish their tales of exploring the French Riviera before settling in Barcelona to complete his medical research. Brett explains,

'Now that we've turned in our rental car, I can focus on my research.'

Brett's so upbeat and happy. He lifts our souls with his uncontainable joy. My body is buoyant as I bounce into my lecture theatre.

Returning home that afternoon, I meet Bonnie in our back yard – her body a limp heap, collapsed onto the grass. Her face pale, stained with tears. Her breathing short and shallow. Her voice frantic. She scarcely gets the words out as she stretches her arms up towards me.

'No, no. Tell me it's not true!' she sobs.

'What? What's happened?'

'Tell me it's not Brett!'

'What's happened to Brett?'

'I can't speak the word. It's too painful!'

'You need to tell me.'

'Brett's dead!'

'Brett who?'

'Our son Brett!'

'No. That can't be! We just talked with him. It can't be true.'

The world seems to crush me. My attempt to stifle my sobbing is hopeless. My gut churns. I choke back vomit. The sorrow seems to poison my blood. I feel nothing in my world matters anymore.

'Why didn't God take me instead of my son?' I wail.

Bonnie and I lie on the grass as a mound of clay, holding each other in unimaginable disbelief. Words do not exist for this kind of anguish. Losing a child is absolutely the worst thing a parent can ever face. I keep thinking, *No it's not true – it can't be.* My life will never be the same again. I absolutely cannot fill that hole.

Bonnie boards a plane for Barcelona within the hour with Jessica's Mother, Ingrid. My heart aches as I kiss Bonnie goodbye in the airport. She's a mess. I'm a mess. I keep hoping, *This can't be true.* My life seems to drain from me as I watch her disappear down the gangway.

It's now up to me to share the horrific news with our extended family – one of the most distressing tasks I've faced in my life. I drive into my in-law's driveway. I sit strapped into my seatbelt, my forehead resting on the steering wheel. I can't move. I want to turn around and drive away. I don't want to go in.

I have no choice. I must go. No one else is here to do it. It's my job. *Why can't someone be announcing my death rather than me proclaiming Brett's?*

I'm greeted at the door by Bonnie's mother, Simone. Immediately she can tell there's bad news. My face cannot disguise my anguish.

'Tell me. What's happened to Shyla?' (Shyla is our dog.)

I blurt it out, 'Brett's dead!'

Simone freezes, unable to respond.

'No! No! It can't be Brett. He can't be dead!' Bonnie's dad, Ken, drops to his knees beside me.

Simone and Ken embrace me as we stand and weep. I can feel our bodies tremble as the news penetrates our souls. We move into the living room. I collapse onto the sofa to start phoning. My body's trembling, my face soaked in tears, my eyes trying to focus on the numbers, my voice breaking – fumbling with what to say. *How do you plan for this?* There is no dress rehearsal. It's beyond belief as I share my insufferable news. I relive the horror again and again as the words tumble from my dry mouth. Darrell, Doug and Gary – Bonnie's brothers. Brett's half-brothers Bryn and Brad Jr; half-sisters Jamey and Jodi. Each call as difficult as the last. Then, my hardest call of the night – to Justin, Brett's brother alone in Golden, Colorado, attending university. Justin and Brett are best friends. I'm so afraid to call him. I agonise over my news.

My fear is overwhelming, but I know I must do it. Am I afraid that I'll break down as I share the news? (I did on every call) Absolutely not. Thank goodness I've learned that my feelings are not to be hidden. My fear is that on every call, I realise I'll never see Brett again (the same feelings I'll have with the death of my other son Brad Jr) – never play, laugh, cry, discuss, learn, hold his children. Yes, I harbour magnificent memories of a relationship with no regrets, but I've been robbed of more father-son events.

***

Bonnie calls daily to share her suffering as the three of them tackle the tasks of having Brett's organs donated, and his body cremated.

'Today was another daunting day. We entered the morgue, asking to see Brett's corpse. I watched the mortician pull back the sheet to uncover Brett's body. I couldn't look. I thought I was going to faint. I bent down to kiss Brett's lifeless face. It was so pale. His sparkling eyes closed. I tried to cut a lock of Brett's hair, but my hand shook violently. I just stood there, stroking his ice-cold face. It was like saying goodbye to a statue. I couldn't bear to hear the words, "Yes, please send his body to be cremated."

'The week never gets easier. Leaving the crematorium carrying Brett's ashes; visiting the park where Brett passed away; packing Brett's belongings; rearranging the flight home'.

'Brett had promised to take Jessica to the beach. Today we lived out his wishes, heading to the sand with his ashes and the mini beer bottles he'd purchased for the celebration. We suntanned amid frolicking children, thong-clad men and topless women – we knew Brett wouldn't miss such a day.'

I keep busy planning for Brett's 'Celebration of Life'. As the three gals fly back into Calgary, we complete the details – we'll meet well-wishers in the church vestibule as they enter to honour Brett. I need the hugs, the eye-to-eye contact, the outpouring of love, the admiration of Brett and the concern for me. It's the only way I can get through the day. More than 1,000 friends honour Brett; Gloria plays and Annabelle sings; family members cover Brett's urn with roses; medical mentors, friends and family share meaningful and comical stories; pastor Ian Trigg shares the meaning of *Reaching the Finish Line* (the thought that none of know when the finish line will appear – we must prepare every day to cross that line victorious).

'Brett lived only twenty-eight years, 340 days. He was in his fourth year of residency when he passed away suddenly while jogging with his wife Jessica in Barcelona. He lived his life to the brim every day in every way – he did more living his twenty-eight years than most people do in a lifetime. His core values, work ethic, integrity, desire for excellence, never-ending pursuit of knowledge and resilience in the face of adversity made Brett such an incredible young doctor with a very promising career ahead of him. Brett was relentless in his work as an orthopaedic surgeon, where his mentors and patients spoke of his professionalism, inconceivable work ethic, incredible knowledge, and genuine compassion. He served as his ego shrank and dedicated himself to healing. He was a relentless instigator of adventure and experience, shining in his motivation of himself and others. He knew when to lead and when to follow, and while he was a trailblazer, he was also a selfless servant to his family, friends, patients and profession.'

Bonnie shares her thoughts.

***

'With Brett's death, the National and International medical community have been robbed of an academic superstar, and perhaps even more poignant than that is the many, many patients who will not be impacted by his compassion, his professionalism and his surgical skills.'

Dr. C. Fisher – Head, Neurosurgical and Orthopaedic Spine Program, Vancouver General Hospital

***

'Brett was a fantastic resident. He would have been a fantastic researcher, and I truly believe that he would have changed how we do things.'

Dr. B. Masri – Head of Orthopaedics, Vancouver General and University Hospitals

***

As Bonnie and I drive home, we're in shock. The grief has hit us like a torpedo. We share how pleased Brett must be as he observes from heaven. 'Always be the best you can be,' he would say. His celebration was indeed as close to being the best it could be. Arriving home, we hug and weep with breath infused with the buttery scent of Fireball, Brett's beverage of choice.

***

Brett was a Chancellor Scholar at the University of Calgary, receiving degrees in Kinesiology and Medicine, and receiving a master's degree from the Harvard School of Medicine. Outside of the classroom, Brett also received U of C 'Dino' athletic scholarships for both volleyball and golf, and enjoyed banging away in the percussion section of the symphonic band.

His insatiable desire to 'be the best he could be' resulted in him achieving the status of twice being a national finalist for the Rhodes Scholarship. Brett's research garnered, posthumously, the prestigious Frank Stinchfield Award for the best research received from a resident or fellow in orthopaedic surgery in North America.

We know Brett would be most humbled that his dreams and compassionate care would be carried forward through the awarding of four significant medical scholarships – one at the University of British Columbia (raising over $300,000), and three more at the University of Calgary. Although Brett's contribution to the medical world had only begun, his scholarships will continue Brett's legacy and unending desire to help others.

Brett's endless pursuit of excellence and innovation is what will live on in all of us who he touched. Every day I ask, 'What would Brett do?'

Brad Jr – My Son

'He's gone! Brad's gone!' Brad Jr's wife Katie's voice is hysterical – barely recognisable. I struggle to understand her.

'He's gone? Where's he gone to?'

I listen intently. I need to understand! Katie's gasping for air between her words. I feel fear and panic in her voice. It's just not like her – she's one of the strongest women I know.

'He's gone! He's gone!'

'Katie, tell me what you mean. Where's he gone to?'

'He's dead!'

'I'll be right over.'

No, how could this be true? Our son Brett had passed away just three years ago. How could I lose another son? Please tell me they've revived Brad Jr.

As Bonnie and I turn the corner, we see the police car and ambulance, lights still flashing. I fling open my door.

'Tell me it's not my son, Bradley.'

'I'm sorry, sir. You should go into the house quickly.'

We burst through the open door. Katie's sobbing. The paramedics are in the bedroom, working on Brad Jr. The police are distracting Cela and Chase, 6-year-old twins. Tayah and Bradyn are still in elementary school. We embrace Katie as we strive to control our emotions. It's impossible – tears soak our cheeks. My body quivers. I don't know what to say. What can I do? It's such a helpless feeling – standing there not able to help Brad Jr as his body lies lifeless on his bed. Katie had tried to resuscitate him with CPR before the paramedics arrived. No luck! It's too much. I can't go through this drill again!

For two years before Brad's death, he suffered from Wegener's disease. He refused to let the disease rule his life. He continued to work and play hard. He never lost his flamboyant personality. Those health challenges made life more difficult in his last months, but he never gave in. He kept fighting until his disease simply overwhelmed him.

Brad Jr was one of the most loving humans I know. His love for Katie and their four kids was the role he cherished most – an amazing scene to witness. His love knew no limits. It was overwhelming. He ensured his family were always safe, protected and unconditionally loved. I too felt his love every time I saw and talked with him. He held nothing back. Tears welled up every time we said goodbye with a bone-crushing hug.

Brad was unbelievably funny – he could have followed a successful career as a stand-up comedian. He was so much fun to be with. He had an uncanny ability to bring laughter wherever he went. He was the silly, goofy dad – the guy who could turn scenarios into laughter – the dad who got his kids excited just before bedtime (much to the chagrin of Katie) – the guy you wanted to sit at the dinner table with. No matter what the situation, he would have us laughing even as we found ourselves in the most ominous event – but not this time. We needed Brad.

Brad was a risk-taker and an adventurer. His entrepreneurial spirit constantly consumed his thinking, looking for new ventures to tackle. He and wife Katie worked their tails off – they were both such hard workers. No task was too much, too dirty, too difficult, or too complicated. They simply rolled up their sleeves and got after it – 'Let's git 'er done.' If something went sideways with their project, he never pointed a finger or blamed. He accepted accountability and went back to work to make it right. Brad was known for his hard work and his charismatic ability to build relationships with colleagues and clients. His reputation was one of reliability and dependability. I'm so proud of him.

An innate curiosity in the world and all it had to offer took him and Katie to the far reaches of the globe. What started off as a vacation to Thailand evolved into multiple businesses importing unique art and furniture pieces into Canadian homes. They put their combined skills and entrepreneurial energy together to start 'Kilbco', a decorative concrete curbing company.

Brad will live on in Katie, his four children and all of us who he touched so deeply. Anyone who sees our smiling faces or hears our joyful laughs will know Brad Jr remains nearby.

REFLECTIONS. Our Creator Will Decide When It's Time

There's not an English word for parents who have lost a child. Nothing had prepared me for the grinding day-to-day pain of a father who'd lost two young sons. Yes, I've sat with my parents as they took their last breaths. It's the way it should happen – children burying their parents. But my biggest fears become reality – living on this planet without Brett and Brad Jr. I must decide how to live life without them – a new norm I never aspired to realise. I knew them as children, as men. I knew their strengths and weaknesses, their loves and dislikes, their joy and tragedies. If you ask me, 'Will I ever see the like of them again?' I would emphatically say, 'No.' Oh, how I wish God would have substituted my life for theirs.

Expressing emotion when going through extreme pain is not weakness – it's raw grief. Each day I endeavour to reflect on the good times I had with my sons – canoeing, climbing, sea kayaking, camping, fishing; on the motocross track, hockey ice, volleyball court, lacrosse rink; international excursions; helping with homework or a science project; talking in our tent, discussing what life really means, what true love is, how to choose a spouse.

They had spent years preparing to contribute back to society when their lives were abruptly cut short. It's so painful to try to understand why these young men died just as they were excelling in their chosen fields. I cannot answer that inexplicable question. I have no answer. They were so accomplished. They were such a light, with so much to give to society. Why not me instead of them? I wanted more time with my boys, to do the things that dads do – like my dad did with me. I could never get enough of them. It really does make us appreciate just how limited our time on earth may be – how little life can be truly planned or accurately anticipated.

I rely on my faith to give me comfort. My soul is strong. I don't know why God snatched my sons away from us. I only know the pain of not having them in my life is immense – a hole that seems so deep and never fills. It sucks! How often must I sob with a salty, wet face?

I experience a peace I cannot explain nor understand. I only know, without doubt, I continue to live with a supernatural peace. I've had a glimpse into the depths of my soul that forced me to validate my faith. And yes, I can unequivocally say I know it's my faith that has made me stronger today.

I have options as I strive to live without them, without Mom and Dad. I've made the decision to keep experiencing and confronting fear head-on; to be more human, more alive, more authentic, more vulnerable, more loving; to find some way to climb back up out of the despair that never goes away; to find personal strength that is stronger than I ever expected; to gain more appreciation and gratitude; to form deeper relationships, focusing on friends with substance; and above all to grow closer to my wife and remaining kids.

Most of us have no say when it will happen. Yes, there are steps we can control in our lives – what we ingest into our bodies, what healthy activities we pursue, how we handle stress, maintaining a positive attitude no matter how dire the circumstances – these will help prolong a healthy, lengthier life. All of us have witnessed relatives or friends who have lived long past expectations, and those who have left us all too early. I must neither focus nor worry about how long I will be here on this planet. Rather, what will I do with my time? How will I live life contributing maximally with the talents I possess? Quality versus quantity. I must live with the trite but true challenge that I throw to my children, athletes, students and myself, 'Live every day to the fullest.'

I can't Imagine processing this grief of my two lost sons on my own. Their lives are woven into the fabric of my heart and soul. I absolutely need my family through my daily grieving – it's an essential part of my coping. As I look into eyes of my wife, kids and grandkids, I see part of Brett and Brad Jr there. It gives me comfort.

John Ross MacRae

MY LOVE AFFAIR WITH FEAR

Brett

Brad Jr

9. Coaching Decisions

'Control the controllable.' **Brad Kilb**

Escaping the Sicilian Mafia

'Are we all here?' I enquire. 'Take a headcount. Push those lockers up against the door.'

The Italian Division One league consists of the twelve top teams. At the end of league play, the bottom two teams in Division One enter a playoff series against the top of two teams in Division Two. The top two teams following this mini four-team playoff series will play in Division One the following season, with the other two dropping to Division Two.

As a pro team, it's extremely important to stay in Division One, since we would lose our million-dollar sponsorship if our team is relegated to Division Two. Our team finishes eleventh in league play, and we must play in this mini playoff. Our playoff series is going well, and we are in the top half of this four-team tourney.

One of the teams is from Sicily. As we fly into Caltagirone, our teammates warn us not to make any jokes about the Mafia. They claim the Sicilian team is sponsored by the Mafia, and stories we hear about Mafia atrocities are true. One recent episode involved an Italian jeweller receiving ransom notes following the kidnapping of their young son. When the money was not forfeited, the Mafia started sending body parts to the parents. Thank goodness we don't have any children yet.

Our match is a do-or-die match for the Sicilians – if they lose this match, they will be relegated to Division Two. We can feel the tension as we enter their competition venue, a converted church with a balcony encircling the court. One of the nasty distractions for our players is the revolting action of fans as they lean over the balcony above our server, spitting into her face as she looks up at the ball. It's perhaps the only time in my coaching career I've not asked my players to 'Keep your eyes on the ball.' The competitive environment is one of the most intense I've ever enjoyed – wild Sicilian fans; must-win opponents; wiping spit from my players' faces; officials paid off by the hosting Mafia. As the final whistle blows, we are jumping up and down with our 3–1 victory.

Our elated feelings of winning are quickly replaced by feelings of fear. The fans are furious. Our victory relegates their team to Division Two. They swarm the court, trying to take out their wrath on us. The police quickly encircle us, arms outstretched, forming a circular human barrier as fans shout profanities, spit on us, launch bottles at us and lash out, attempting to kick and punch us. Mob mentality spreads like wildfire as the enraged fans try to get at us. I'm terrified – so grateful for our police protection.

The gendarmes struggle to escort us to our dressing room. It's bedlam. They instruct us to lock the door and stay inside. We can hear the police shouting as the mob scuffles outside the door. For now, we're safely barricaded.

Once inside our fortified dressing room, I start planning our next move. It's too dangerous to leave and go back to our hotel. I decide we'll wait until 5:00 a.m., flee to our hotel, pick up our belongings, and 'get out of Dodge'.

Our escape plan works perfectly. We finally flash our victory smiles as our plane lifts off. Another playoff win. We will stay in Division One next season. The headlines in the Italian newspaper label Bonnie and I, 'Salvatore' – Saviours.

Undercover Agents

As I prepare to leave for the gym for our friendly volleyball match, Isabella, our Polish guide comes bursting into my room with tears rolling down her cheeks.

'You must leave Poland tomorrow! You are not allowed to stay in my country any longer!'

We are on the fifth day of our twenty-one-day training and competition tour with the University of Calgary women's volleyball team in 1987. My development programme as a head coach has always included international tours. I believe such tours not only improve our volleyball skills, but also introduce us to a different culture as we travel the host country, integrating volleyball, culture and sightseeing. A normal day for us typically includes three activities – a training session with a Polish coach and players, a cultural excursion and a competitive match against the Polish team we train with in the morning. I always look forward to the training sessions since it's not only my players who improve their skills, but I'm also exposed to numerous different coaching pedagogies. Today we've followed our typical routine.

Why are we being forced to leave Poland? Why is our pre-paid sport tour to be cut short? What have I done to warrant expulsion from the country? Isabella, a Polish university student, explains.

This morning following our practice session with the Polish team, I asked the coach if I could meet with him to share volleyball techniques. He readily agreed and invited me to his home to spend a couple of hours discussing volleyball. I didn't realise he was an anti-government protestor – not a member of the Communist Party. Our guide continues to explain that a KGB agent had been assigned to follow our team and observe all our interactions. My morning visit with a non-sympathiser with the communist philosophy triggered the KGB agent to report, 'They're not a volleyball team, but Western spies engaged in surveillance.'

Isabella continues to explain the decision is irrevocable, and we must pack our bags this evening, following the match. I'm dumbfounded. I feel sick. I don't know what to do. I didn't realise my mistake visiting the home of an anti-communist coach would trigger such an outcome.

Sitting in my hotel room with our sobbing guide, I struggle to come up with a game plan that will allow us to finish our three-week tour. *What can I do to change the mind of this agent? Is there any way I can change his opinion?*

I finally come up with a plan. I ask Isabella to rank our two countries. Who is the stronger country in terms of international volleyball? The obvious answer is Poland. I explain that's why I've brought my team to

Poland – to experience Polish volleyball and learn to become a better coach. I reason with Isabella and ask her to go and present my proposition.

At this morning's training session with the team we are to compete against this evening, I've observed the Polish players very carefully. I believe we can beat this team in the 'friendly' match this evening. It's a big gamble – not so friendly, but I'm willing to take it.

'Go and tell the agent I understand Poland is a stronger volleyball nation than Canada. If we are successful in defeating the Polish team tonight, it will prove to him that we *are* volleyball players, and we shall be allowed to stay in Poland until the end of our tour. If we lose, we will pack our bags and leave the country tomorrow.'

As Isabella departs to explain my proposition to the agent, my heart pounds with fear as I wonder what I've just agreed to. *What will I say to my players as we pack our bags? Will the Polish media label me a spy posing as a coach? Will I be thrown into jail as my players fly home?*

It's now time to focus on my volleyball game plan to ensure victory. In our pregame session with our team, I try to stay relaxed despite my gut churning with butterflies. If we lose tonight, it will be a very expensive loss. I hope and pray my confidence is to be rewarded.

Brad, focus on the moment. On w*hat we must do to win this match. What technical information can I give the girls? What weaknesses have I detected that we can exploit?*

As the match progresses, my players are executing my game plan perfectly. We lead the best of five match 2-1. My brain wanders. *Just one more set win, and we stay. We've got to win this crucial next set. I don't want to go to five sets.*

We score the last point. The set and match or ours, 3–1. You cannot imagine my relief and joy as I meet my team in our post-game debrief. It's only now that I share with my coaches and players my gamble – my brazen confidence in my team. The room falls silent as my players look at each other in disbelief – and then the cheering erupts. My faith in our team has paid off, big time. My gut feeling told me to make this proposal, and it proved right. I'm so glad I never placed that stress on my players before we played.

We get to spend the next sixteen days in Poland without any form of harassment. The only teasing I receive is from my own players, who could not believe I'd put so much on the line based on their skill and determination. Some would call me a risk-taker. Some would say I'm overconfident. I would say, 'Embrace the fear as you follow your gut feeling based on previous experience.'

Double or Nothing

I'm in Hawaii with my University of Calgary women's volleyball team. As the head coach, I've arranged an exhibition tour with several universities on two islands here in the Pacific. Hawaii is a hotbed of women's volleyball, and I want my team to play some of the top-ranked teams in the US. I must admit that travelling to Hawaii on a business trip is not such a bad gig.

An international tour like this is certainly expensive. However, through my international contacts, I've been able to arrange guarantees to play these Hawaiian teams. Each team we play will pay us $2,000 USD per five-set match. It's not unusual for a Hawaiian university to offer a guarantee to top teams to come over and play, since for them to travel to the mainland is very expensive. I've been able to arrange six matches, which helps to offset our $20,000 expense.

I've purposely scheduled the first day to simulate some unusual travel arrangements. I remember chatting with my dad about peaking at the right time of the season. 'Your win-loss record going into the playoffs is meaningless. Don't be afraid to swallow some losses to prepare for the important part of the season.'

My goals with exhibition matches are always long-term. Winning these matches is not as important as preparing my team for future league and playoff matches. I want my players to learn some lessons, perhaps the hard way. We leave Calgary on an early morning flight, arrive in the heat of Hawaii mid-day, check into our hotel, go for a jog in the sweltering sun, eat our pregame meal and head directly for our first evening match against the University of Hawaii, Hilo. Our results are not as I had planned. We lose all five sets. A miserable showing below the calibre of our team.

Following the match, the disgruntled athletic director makes a beeline for me. He explains how disappointed he is that his coach had arranged to play us with a guarantee. He rambles on about the low calibre of our team, and he's not looking forward to playing us once again the following night. He even goes as far to say, 'Your team is an embarrassment. I really don't feel you've earned your guarantee tonight.'

I can understand his dilemma, so I make him a proposal.

'You don't have to pay us the guarantee for tonight's loss, nor for tomorrow night. However, if we beat your team tomorrow night, you must pay us double the guarantee for both matches.'

He looks at me with delight written all over his face.

'Are you telling me that when we beat you tomorrow night, we will owe you nothing?'

'That's right. It's either $0 or $8,000 – it's your choice.'

'I can't believe you'll agree to this.'

We shake on our new arrangement. As he puts the envelope back into his suit pocket, fear strikes. I wonder if I've been overconfident in suggesting this proposal. It's not the first time I've made a mistake! Thank goodness we have already paid for our return flights – I don't want to row home!

Back in the hotel, I share the lessons I hope we are learning from this hectic travel day. We discuss the loss and the challenges we are attempting to overcome. I don't mention anything about the guarantee to the players, but I throw out a challenge to them about coming back after a devastating loss and learning how to be resilient – playing up to our level following a disastrous loss. I secretly realise that rising to meet this challenge could make a big difference in their confidence, and my pocketbook!

Today during our morning pre-competition training session, I can feel an air of excitement and optimism within our players. I'm hoping I'm reading the players correctly, since last night I really didn't get much sleep. A $4,000 deficit swing in our budget would be disastrous. As we leave the venue following our training session, I pick up a local newspaper only to discover the headlines on the sports page, 'Calgary Coach Guarantees a Win.'

The article goes on to explain how UH had dominated their Canadian opponents, and how I have guaranteed a victory tonight. Well, nothing like adding pressure to my already heightened anxiety. In our pre-match meeting, I pull out the article and post it on the bulletin board. I explain that my intent is not to add pressure, but I want them to understand how confident I am in their ability to play our best tonight. I emphasise the importance of taking care of business on our side of the net rather than focusing on the loss.

Before we depart from our van, I throw out one more challenge.

'Enter this gym tonight with your heads high. Believe you can play your best. Strut in with a confident swagger.'

As I glance around the sold-out gym, I can't help but notice the over-confident look on the UH coach. I see the UH athletic director (AD) sitting in the front row with a smile, reminding me of a trader whose stock has just skyrocketed. As I make eye contact with the AD, I say to myself, *I hope you've got that eight grand in your pocket,* followed by a short prayer. As the match ends and we shake hands, the AD walks over to me, reaches into his pocket and pulls out an envelope.

'I can't believe the turnaround in your team. I would never have guessed this outcome,' he growls.

'That's sport. Thanks for honouring our agreement.'

I feel a huge relief and charge over to my team with animated high fives. They still have no idea of the financial significance of this match. I know they're delighted with meeting my challenge and the comeback with a match they can be proud of. In the post-game meeting, I explain the gamble I made, and the celebration begins – whoops, hollers, high fives and hugs. They are overcome with joy as they comprehend the confidence I have in them as a team. Mission accomplished – we've learned what it means to be resilient as I treat them to a lavish post-game meal.

Casino Float

'Our take-home net profit is $65,000,' the announcement comes from our volleyball club president following our two-day casino.

In Alberta, the Gaming Commission has a scheme whereby non-profit organisations can volunteer members to work casinos in conjunction with professional employees. The agreement is that our club provides thirty-three volunteer shifts, lasting from noon until 2:00 a.m. Our club then splits the profit with the casino owner. Our cut of the profit is the main source of our annual operating budget, allowing us to do unique training and competitions with our boys.

One of the requirements of the casino operation is that our club must deliver a $250,000 cash float in specific denominations to the casino. The float is returned the day following our two-day casino. Most organisations pay for a bank loan and armoured truck delivery. These costs are paid from the profits.

After discussions with Bonnie, we decide to play the role of bank and armoured truck. We will take out a personal loan and make the delivery of the cash to and from the casino site. Bonnie begrudgingly agrees but refuses to accompany me on the deliveries in the 'shotgun' position. We'll charge our club the same amount equal to those of the bank and armoured truck company.

I call our bank, 'I'd like to arrange a short-term loan for $250,000, with our house as collateral.'

'Yes. Brad, where would you like to have it delivered?'

'I'll pick it up personally on Wednesday.'

'I don't believe that would be a good idea. Don't you remember that a casino owner was recently held up and shot as he delivered cash to his bank?'

'Yes, I read about that in the newspaper. However, I'll pick up and deliver the cash myself.'

'We'll have the cash ready for you.'

I'm escorted to a private room in the bank. The door's locked, curtains pulled and we begin to count the money. Wow. That's a lot of cash! I stuff it into my briefcase and prepare to leave.

'I'm sorry. Normally we would escort you to your vehicle with your cash withdrawal, but it's too dangerous with that amount. You're on your own, Brad.'

As I carry the wad of cash in my briefcase, my feelings fluctuate between the fear of getting shot, and the excitement of holding such a large chunk of cash. *Is this wise? Should I wander through the parking lot with this much cash? Is the outcome worth the risk?* Suddenly, my fear grows. I'm scared.

How can I walk nonchalantly out the door clutching $250,000? Should I be carrying a gun? I pull the delivery off successfully without dodging bullets.

Motocross Nationals

It's a dream for so many boys as they grow up.

'Dad, can I get a motorcycle?' Brett is ten years old, very athletic, an extremely hard worker who never gives up as he pursues excellence.

'If Brett gets one, I want one too.' Justin is eight years old, an extremely gifted athlete. It doesn't matter what sport he takes up – in an amazingly short time, he becomes adept.

Both boys have excelled in many of the traditional sports – ice hockey, soccer, basketball, wrestling, track, badminton, alpine skiing. Perhaps it's time they try an extreme sport such as motocross – that crazy sport where riders spend as much time in the air as they do on the track. Handling jumps seems to come naturally to our boys, since they build bike jumps in our backyard. Although we encourage them to challenge themselves, we often wonder about our decision as we sit in the Children's Hospital Emergency Ward following another motocross mishap. Motocross is an activity where you either win, or crash – to be successful, you must always ride on the edge of disaster.

Coming second would be satisfying to many competitors, but not with our boys. We share our desire for them to simply 'be the best they can be', win or lose. But they only compete to win. Every time the start gate drops, the fear of a life-altering tragedy clouds my cheering. As the chequered flag comes out, I breathe a sigh of relief.

It helps that Bonnie's brother Doug owns Blackfoot Motorsports. We find ourselves with Doug, purchasing motocross racing bikes, all the protective gear and building a trailer for our boys.

Although I owned and rode a 650cc motorcycle as a student in Africa, I'm not a motocross competitor, but my coaching background takes over as I trailer the boys to the track for practice sessions four times a week. I don't know much about the technique of racing, but I understand the framework of training athletes.

'All right, if you boys are going to race, you're going to be competitive,' I insist. 'We'll start every day on the track practicing the rudiments of successful racing.'

We have a regimented training session working on different aspects of motocross – starts, high and low berm turns, jumps, seat bounce, whoops, tabletops and when to 'pin it'. Both boys adopt the regime, work hard and improve quickly.

I feel inadequate when it comes to the technique of competitive riding. My ability to give error detection and correction was very low, particularly as the boys progress very rapidly and began winning. The only principle I really understand is, 'You're either on the gas or on the brake – there's no coasting!'

As the boys progress, the skill set becomes more and more difficult. I feel it's time for Brett to work on a triple jump. I truly believe he's physically and psychologically ready to master this challenge. A triple jump is much faster than taking each jump individually, but if you don't make the full jump, clearing all three, you will suffer a horrific crash.

'Brett, I know you can make it,' I confidently declare.

Brett revs his engine, accelerates quickly down the track, launches into the air on the first jump and soars towards the third.

His bike falls short, crashing into the front side of the third jump as he's jettisoned out the 'front door' of his bike. He has taken his hand off full throttle at the last minute! He lays unconscious on the track.

I'm so afraid. I feel so badly. I've let him down. *How can he trust me in the future? How should I proceed as his coach? Will he be willing to try a triple again?* I hate that feeling of incompetency. *Why did I ever agree to coach my son?* As Justin and I rush over to Brett's limp body slumped on the track, I say a prayer.

Brett eventually sits up. 'Dad, can I try that again? I know I can make it.'

Try it again he does, this time successfully. He's that kind of kid. He never gives up.

***

Justin and Brett are doing extremely well in our weekly local and provincial motocross races. Brett races in the 60cc class, and Justin in the 50. It's time to head to British Columbia to test the boys in the 1998 National Motocross Championships. We arrive in Abbotsford in our homemade trailer, unload our bikes and roll out our sleeping bags on the floor of our multi-purpose trailer. It's so intimidating to look around the campsite at the fifth-wheel trailers, complete with living quarters and mechanic workshops, from across Canada. Many of the boys have a race mechanic who travels with them to every race. I certainly do not qualify as a mechanic. If anything goes awry with the bikes, I'm sunk! Our makeshift pit gets numerous perplexed looks from other competitors, along with a few giggles. Fear floods my brain. *Is this national competition too elite for my boys? Am I creating a scenario of failure?*

Although there are many outstanding racers in the boys' age categories from across the nation, the talk in the pits is all about two boys from British Columbia who have been tearing the track up all season. The betting for a national champion is heavily weighted in favour of one of these two. The championship will be decided by the total number of points scored through three heats.

Justin's 50cc category is an exhibition race for young riders. There's no national champion declared. Justin races extremely well all weekend and finishes on the podium.

As Brett lines up in the start gate for the first of three motos, the announcer blares out, 'Here we are at the national championships. It's going to be a dogfight between Tommy and Johnny, the two superstars from British Columbia. We'll have to wait to see who wins this dual battle and takes home the national championship trophy. Let's see who gets that all-important holeshot in this first moto.'

I'm hoping Brett does not hear the confident announcer as he promotes his hometown favourites.

'And they explode out of the start gate,' barks the announcer. 'Who's got that holeshot? What, it's neither of the favourites! Let me check his number to see who this rider is.'

'It's a kid from Alberta racing for Blackfoot Racing. We'll see how long it takes for him to be overtaken by Tommy and Johnny.'

'There's the white flag indicating the final lap of this moto. The Blackfoot racer is still in first place, hotly pursued by Tommy and Johnny.'

'And today we've seen an amazing upset. Kilb, from Alberta, has won this first moto and collects the highest number of points in this national championship. We'll see if he can hang onto that prestigious position tomorrow.'

Brett, Justin and I head to the wash rack to wash down their bikes. We're bombarded with questions from the dads of other riders.

'What kind of fuel are you using? Is your kid racing in the right age bracket? Where did he learn to ride like that?'

I could feel the tension from these BC parents.

The second moto is an exact copy of the first. Brett finishes first and makes his victory lap, carrying the chequered flag. Back in our pit, we're ecstatic. Whoever thought after two motos Brett would be sitting comfortably in first place? The BC contingent is not happy.

As we prepare for the third and final moto, I sit down with Brett for our pre-race talk.

'Brett, you realise you're comfortably sitting in first place as you enter this final moto. Those two BC boys are chomping at the bit and want to fight it out to win at least one moto before heading home. I want you to realise that even if you finish in third place, you'll be crowned the national champion. Don't get mixed up with those two. If you crash, you'll lose the championship. You can just hang back, finish third, and go home with the national trophy.'

'Dad, are you asking me to finish third?'

'Don't you understand? If you finish third, you'll be crowned the national champion. If you crash, you'll go home empty-handed. Finishing third in this moto is not a bad choice.'

'But Dad, you've always urged me to be the best I can be. Finishing third intentionally is not being the best I can be. I'm not sure I can do that.'

I terrified, but what else could I expect? Our whole life together has been striving to be the best we can be at everything we do. Brett's right, my bad. I'm working out the numbers game, but he's consumed by the emotion of competing. I'm looking at the end result – Brett standing atop the podium, while Brett's focusing on the moment. I'm envisioning myself strutting around as the proud parent of a national champion, while he's concentrating on winning another moto.

Brett explodes out of the starting gate, wins the holeshot, leads every lap of the entire race, and once again makes his victory lap. He's the national champion. I can't express how proud I am of Brett, especially since he refused to listen to my advice.

Those of us engaged in motor sport understand that success is dependent upon both rider and machine. Brett's championship could not have happened without Justin's involvement. Justin had just received a new 60cc bike weeks before the championships. On our trip out to BC, the three of us discussed Brett using Justin's bike for the championships, with Justin racing the 50. Justin displays that rare combination of elitism, humility and empathy. He's willing to work quietly in the background while others receive the accolades. It's so evident as he plays elite volleyball – setting up the attacker who receives all the glory. Justin agrees and Brett goes on to win on Justin's bike. I struggle with this decision to this day. Had I not suggested to let Brett ride Justin's bike, there would be no championship trophy sitting in our house – 50cc does not declare a national champion. I'm not sure that Justin doesn't despair his chance of winning the 60cc championship. As their coach, I made my decision based on previous competitions.

Sometimes There's a Cost to Doing the Right Thing

My volleyball coaching career began as I coached junior and senior high teams in the school where I was the physical educator in 1972. I had never played the game, but I chose coaching volleyball since I was teaching in a very small school. I also felt that volleyball was a relatively easy sport since you simply touch the ball three times and put it over the net into the opponent's court, a space big enough to land a helicopter.

My coaching learning curve spiralled quickly upwards as I grabbed every opportunity available to learn about the sport – shadowing one of the most successful volleyball coaches in the city, inviting guest coaches to our training sessions; spending time with Olympians and reading as much as I could.

One summer, I brought my team together for an intensive one-week skill development camp. Striving to improve my own coaching skills, I brought in an Olympian, Debbie, to lead the camp. Shortly after the camp finished, Debbie phoned me to ask if I would coach her team – one of the top two women's teams in Canada. I hung up, thinking she was teasing me since I had so little experience.

A couple of days later, she called again. 'Brad, we would like you to coach our team. Don't hang up.'

'Debbie, you know I'm not qualified to coach such an elite team.'

'We have the expertise to look after the technical aspects. You are a master motivator, and that's why we want you to coach us.'

The fear of accepting a role we don't feel qualified for is something many of us face. It's certainly the way I feel as I receive this invitation. *I'm not the person who should be coaching one of the top teams in Canada.* I look at my weaknesses rather than my strengths. I fail to see my value in the eyes of this team, a value they require. What a lesson for me to learn – to acknowledge my fear, but to lean into a challenge utilising my gifts while learning new, complementary talents.

I accept the position and it works out extremely well. The outstanding players on our team, led by captain Debbie, assist me in so many ways to understand the finer technical points of coaching. The players use their expertise to win the silver medal at the Canadian National Open Championships. Those elite players not only assisted me in my growth as a coach, but also as a more confident human being. With this stretching encounter, our school became known as a volleyball powerhouse as we won Division, Provincial, and Western Canadian Championships.

***

Our women's high school team is breezing along in our regional competition. We have sewn up first place and are preparing for playoffs. My team captain asks if we could meet.

'I'm not sure if you're aware, but one of our players is taking drugs,' she claims.

'I have my suspicions, but I have no proof.'

'Well, every player on your team could confirm it.'

My dilemma is not simply addressed. Our school has a zero-tolerance policy for drugs, resulting in expulsion. If I don't act, I will lose credibility with my team. Word will spread throughout the school, and my integrity would sink quickly to zero. I must act, but I'm afraid. Afraid for the player since 'being cut' from a team can be devastating – often, teammates are your best friends! And afraid for myself – this player is the daughter of our school principal. What a disaster. How embarrassing. *What should I do? The principal has the power to fire me on the spot. Is my job security more important than my integrity?*

I meet with the player before practice.

'I'm sorry, but I'll have to release you for contravening school policy,' I state.

'Are you really sure you want to do this? Do you remember who my father is?'

'I'm sorry. Please bring in your uniform tomorrow.'

I must risk my job to report the truth. Following practice, I enter the principal's office.

'I really don't want to tell you this, but I must. I released your daughter from our volleyball team today for drug use.'

'Ohhhhh! Damn! My worst nightmare,' he responds.

'I feel her punishment has been sufficient. I don't believe you should expel her from school. Why she has been released from the volleyball team is between you and me. I'm not going to mention why I released her to anyone. The decision of what happens now is up to you. I will support you with whatever decision you make.'

The fear of making such a stand seems overwhelming – terrifying. I know what I did is a trait my father always preached and lived by, 'Make sure you do what is right, not what may seem to be the easiest.' As a nine-year-old, I saw my father ready to be cut off from his parents by the decision he made regarding my adopted sister. Today, I am ready to lose my job by making the right decision.

World Championships, Rio de Janeiro

It's my first World Volleyball Championships as a head coach. My team is the Canadian National Junior Women's Team, nineteen and under, and it's our first worlds in 1977. Normally on the first day of training, we declare our training venue 'Closed'. No one is allowed in the venue except for the team training. Since none of us have seen any other teams play, it's important this first training session is not scouted by opponent coaches. Argentina, the team we will play this evening, is training in the same venue immediately following us.

Our training goes well as we practice some of the plays we will use against Argentina. As our training session ends, I call my assistant coach over.

'You take the team back to the hotel. I'll find my own way back.'

Before the Argentine team enters, I saunter upstairs to a janitor's storage room. Once inside, I pull on a janitor's coveralls and hat, grab a broom and start sweeping the stands. The Argentine team carries on their training, unaware the Canadian coach is watching every action. I focus on the team's weaknesses to exploit, and their strengths to counter with a good matchup. I leave the venue with some notes. I must admit – it's the only time I remember being happy to be a janitor.

OK, I'm sure you're asking if this is 'gamesmanship', or 'cheating'. If you say cheating, I agree. *How could I cheat? Can I say it's acceptable because it's an international competition which is loaded with unethical actions?* Personally, I'm not ready to accept any of these excuses. I've coached at every level imaginable, from community grassroots to professional levels all over the globe. I know it's not right just because everyone else does it. As an official at the 2012 London Olympics, I watched numerous unethical actions. As a coach I question, *what I should do to make sure the playing field is equal.* No one should have an advantage due to cheating, regardless of the level.

What I did was wrong. I should not have done it. I'm embarrassed to say I'm a cheater, even though I don't believe it altered the outcome of our match. *Brad, enough of these unethical actions – it's not my way.*

REFLECTIONS. There's More to Sport Than the Score

Let me share a story about my coaching which I hope will shine a light on my philosophy of coaching. I'd call it 'prostituting my values'.

Our University of Calgary Division 1 Women's varsity volleyball team has been blessed with success, riding on the talents of very gifted and focused student-athletes. As the head coach in my thirteenth season, we have consistently been ranked in the top-four nationally, earning our way into the National University Championship Finals numerous times, coming away with too many silver medals. For many coaches, a national silver medal would be most satisfying. For me, not so. I crave that first national championship ring. Yes, I've won the honour of being selected 'Canadian University Coach of the Year', and 'Conference Coach of the Year' four times, but it's the team victory I strive for.

We enter the national championship tournament with seven other teams from across the country. We'll play in the quarterfinals, semifinals, and gold-medal matches. The top two ranked teams have been perennial powerhouse programmes – our own conference champion who we lost to in our conference championships, and the five-time reigning national champions. We've lost to both these teams all season. As a wildcard entry due to our high ranking nationally, we enter the tournament as a dark horse, ranked fifth. Preparation has been intentional and specific. We've prepared all year with these two teams in our sights. It's time to shed the bridesmaid veil and step away with a gold medal. As the coach, I enter the tournament with confidence – peaking at the right time, ready for these top teams.

The word on the street is that I have gifted athletes, but I'm not tough enough as their coach – they say mental toughness is something my players lack. As I compare myself to the reigning national champion coach, I wonder if that reputation isn't true. I've always considered myself a sound technical and motivational coach. However, I know I'm not as hard on my athletes as perhaps I should be. My values and philosophy accentuate the value of each player as they contribute to the success of the team. My coaching style is positive, with instructional feedback.

Having observed the reigning national champion coach, I realise his style is very different than mine, bordering on the edge of being abusive both physically and mentally. Coveting the gold medal that has slipped through my hands so many times, I decide I will abandon my coaching style and mimic the champion coach. A I look back on that season of training, realising it was such a drastic change of character, I ask myself, *Why was I willing to do this?* I believe the root of it was fear – fear of embarrassment, of not reaching my coaching potential, of not preparing my athletes psychologically, of losing 'the big one'. My ego got in the way! My goal became more important than my character! My reputation reigned supreme!

All season I step out of my comfort zone in an attempt to develop mental toughness within my players. I adopt a drastically different coach's methodology to be ready to walk away with gold. Pushing them beyond their physical, emotional and psychological limits; playing 'head games' with them; forcing them to do things they don't want to do. It's not fun. Neither my players, coaching staff nor I enjoy that season of training. It's completely foreign to me, my values and my positive coaching style. I'm a different person as I enter the gym – someone neither my team nor I recognise.

Our three-day tourney starts with a quarterfinal win over the fourth-ranked team in straight sets, 3–0. Perhaps we are peaking at the right time? In the semis, we meet the defending national champions. Our semi ends with an unexpected result, a 3–1 victory for us. The defending champions first playoff loss in six years. Wow! What an accomplishment. Perhaps I was right in changing my coaching style?

The final match is a battle with our conference champions, the team we lost to one week earlier in our conference championships. This weekend is different, very different. Not only have we beaten the national champions, but we come out of the finals 3–0 winners. We are national champions. I raise the trophy, hang my gold medal around my neck, high five and hug my players and coaches. My euphoric emotions are overwhelming. My fear of not being recognised evaporates. My fear of never being a national champion is gone. My emotions seem to be those of relief, rather than victory.

I strut from the gym back to my hotel room alone, pondering what has just transpired. I carry the trophy in one hand and the championship banner in the other. *Is this real? Have we actually accomplished our goal? Will my naysayers now stop criticising?*

MY LOVE AFFAIR WITH FEAR

I thrust open my hotel room door, slam it shut and throw the trophy and banner onto my bed. I sit hunched over, head cradled in my hands, and sob uncontrollably. I weep because I am human. Strangely, my tears are not those of joy but rather of disappointment, of regret, of disgust, of sorrow. I've prostituted my values. I've abandoned my philosophy. I've coached in a style so foreign and repulsive to me. My heart aches. *Why did I do it? Why was winning more important than my values? Would my players ever forgive me?* I sit in a pool of tears as I hear a knock on my door. No, I am not answering. I cannot face anybody in this state of mind. I should be jumping around with joy.

Yes, I got my gold medal, my championship ring and an elevated reputation. But it was *not* worth it. I've never coached that way again, although now I have more championship rings than I have fingers.

I believe most coaches can be divided into two categories – those who are there to use their athletes to enhance their own reputation and status; and those who are there to assist in the development of their players' expertise and character by benching their own ego. Our coaching legacy is founded on our core values. My core values, based on my Christian faith, will determine the decisions I make, the way I behave and how I will treat my athletes, coaching staff, opponents and officials. Those values will determine my vision statement and coaching philosophy, 'Winning will never again come at the expense of my personal values.'

***

Don't let anyone fool you – coaching is a very difficult role. We must establish trusting relationships with athletes and parents; make decisive decisions, perhaps some that not all agree with; be analytically sound, technically and tactically; focus on solutions, not only problems; ensure consistency in areas such as discipline and commitment; be demanding as we push athletes beyond their comfort zone; and hold athletes accountable.

It's essential for me to share who I am, my coaching vision, values and expectations, before the season starts. One of my fears is that once into the season, I'll have to release an athlete who never understood the culture they were buying in to.

Values
- To be true to my Christian values – do nothing that demeans or dishonours myself, my athletes, my coaches.
- To share my belief that family is the most important institution in the lives of my athletes, then education, and then sport.

Culture
- To intentionally curate a culture reflecting my core values.
- To identify a sense of vision and shared goals, with buy-in by my athletes and coaches.

Relational
- To build a trust relationship with my athletes and coaching staff.
- To foster a degree of camaraderie that maybe only wartime soldiering can surpass.

Technical Skills
- To teach progressively and sequentially, assuring every player has sound fundamentals to build and showcase their genius.
- To be decisive in my explanations and feedback.
- To strive for excellence in execution.
- To place each athlete and coach in a position to fully develop and utilise their unique gifts for the success of the team.
- To be willing to experiment. If it doesn't work, throw it out. If it works, put it in the toolbox.

Demanding
- To be directive and decisive.
- To pay attention to what I see vs. ignore, what I celebrate and the non-verbal messages I display.

Specific Role
- To give every athlete and coach the opportunity to fulfil a specific role, leading to the success of our team.

Solution Focused
- To focus on the process of individual and team development rather than on the outcome.

- To develop a spirit of joyful competition – a never-quit attitude.
- To create a structured learning environment with analytical tenacity.

Expectations
- To work together on and off the court, developing behaviours that will demonstrate growth in character.
- To encourage athletes to contribute their talent for the joint success of the team, rather than self-enhancement.
- To teach a realistic level of risk, be willing to experiment, succeed or fail and grow.
- To develop team systems around individual talents, rather than forcing players into a system.

Balance
- To strike a balance between discipline and creativity; success and growth; focus and fun; family, academics, and sport.

Elite West, USA Club Champs

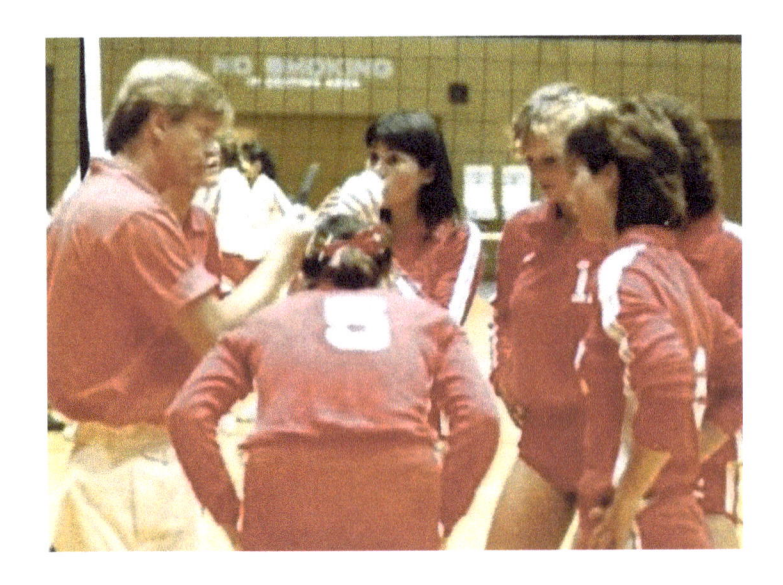

University of Calgary

2012 Olympics

Bonnie and Brad Coaching

Bonnie Coaching with Brett

Justin

Brett, National Motocross Champ

Justin & Brett

10. Unpredictable Animals

'I travel because I'd rather look back at my life, saying "I can't believe I did that", instead of "If only I had."' **Florine Bos**

Italian Cinghiale

'If he charges, he could kill!' I whimper.

My coaching buddy Antonio and I are staring down a ferocious wild boar. Although our plan is working and the boar has stopped running, I'm second-guessing our choice of tactics. Realising he's boxed into the corner with no escape, he's now turned to face us. OMG, what now! It's a standoff at the cinghiale corral. The snarling, grunting boar stares us down – tusks bared, mouth frothing, the hair on his neck erect, pawing the earth like a rodeo bull. Antonio and I stand side-by-side in crouched, defensive positions, waiting for the boar to make his move. My brain is in overdrive, filled with fear. *When he charges, do I jump aside or engage in full contact?*

We're up in the mountains in Tuscany, Italy, as Bonnie and I spend a year playing and coaching professional volleyball. Antonio's invited us to participate in an annual local festival. Always looking for a new adventure, we agree to join him. Antonio is the Italian national volleyball coach, and his girlfriend Gloria plays on the same pro team as Bonnie – the team I'm coaching. As we arrive in the hilltop Tuscan village, we can feel the electric atmosphere. The townsfolk are fully engaged, and the celebratory environment is contagious. We can tell this is going to be a memorable day.

Although we're aware the festival centres around the celebration of the Italian wild boar, the *cinghiale*, I'm not aware what my participation is to be. Antonio describes the festival – dancing in the city square; enjoying barbecued wild boar; buying memorabilia of the wild boar, including menacing ivory tusks. We park and jostle through the crowded, narrow streets towards the registration table. As we arrive, I notice the schoolyard perimeter is completely encircled with two-metre, reinforced steel-wire fencing. Crowds have gathered around the fence to ensure a clear view of the event. Antonio completes the registration and joins us.

'What's the fencing for?'

'Oh, didn't I tell you, Brad? That's what we just signed up for. It's called the "Cinghiale Wrestle".'

'Are you kidding me? We're going into that ring with a wild boar? You can't be serious. How big is this feral beast?'

'Let's go over to the pen and have a look at the critter we'll be capturing. I'm sure it couldn't be that big.'

We stroll over to the cage. I hear snorting and grunting. My pace slows a little. *What have I signed up for? I'm in Italy to coach, not to get gored by a wild boar!* I can tell by the distraught look on those peering into the cage this is no piglet. As I look at the ferocious, mud-covered boar and out at the fenced field, my mind quickly fills with visions of the Roman coliseum. I know the quote says, 'When in Rome do as the Romans do', but …

'He's a little bigger than I thought. Easier to grab and get to the ground.' Antonio sounds almost convincing.

'Are we allowed to use equipment? Do they supply nets or ropes? Are we still able to withdraw?'

Antonio gives me a look of disdain. I fight to hide my fear. It's one of those moments when it seems the odds are against us. When deciding whether or not to jump into the ring, I try to make an intelligent decision. What are the odds of surviving unscathed? Do I have the skill set to be successful? This is one of those escapades where victory seems unachievable. But I've committed to joining Antonio, so I decide to persist and manage my fear as I focus on the task.

Our contest consists of two-person teams – eager weekend warriors lacking brainpower. Each team has a chance to pursue, catch and wrestle the boar to the ground within a time limit. The boar is released with a three-metre head start, and the pursuit is on. If more than one team can accomplish this harrowing task, the winner will be declared by the shortest time. Bonnie gives me a glare of disbelief, followed by words of advice. 'You'd better warm up.'

The PA system bellows, 'Taking the ring is the elite and fearless coaching team from Cecina, Antonio Jacobi and Brad Kilb.'

I had hoped they had lost our registration form, but no such luck … and where did they get that false impression, 'fearless'? There's no turning back. I enter the ring with weak knees, a pounding heart and fear filling my still intact cranium.

They release the boar, and the clock is running. I'm hard-pressed to say my pursuit could be classified as a sprint. I'm not sure if it was my weak knees or my fearful brain that was slowing me down. I find it difficult to keep up with Antonio, let alone the charging hog.

The pig seems to be slowing down, obviously not used to being the prey of crazy humans in front of madly cheering fans. As we close the gap, I get a clear look at the boar's face and tusks. *Wow, those tusks could shred my body and tear me apart in seconds! What I am I doing trying to wrestle this hairy beast to the ground?*

I'm happy Antonio is leading as we get to within a metre of the scrambling boar – he's much closer to those menacing tusks than I. We both make a headlong dive and make contact with the hairy, coarse hide. An ear-piercing squeal fills the playground … the boar escapes our grasp. I'm not sure whether to be happy or sad as I tumble across the dirt, clumps of boar hide in my sweaty palms, a mouthful of dust, the undeniable stench of pig filling my nostrils. Surprised I'm still alive, I scan my body for wounds.

'Let's go. He's getting away. Let's corner him,' shouts Antonio. 'We've got him now. Let's not let him slip through our hands this time.'

I can feel the pressure. I can't let my teammate down. Many a weekend back in Alberta would find me competing in a rodeo 'wild horse race' – but horses don't threaten with menacing tusks. *I've got to find the courage to bear down and do my job.* My hands are shaking, my legs feel like shredded pork, my throat's dry. I'm wishing I could wake up from this nightmare.

Just then, the horn sounds indicating we've reached our time limit without capturing the boar. Game over. The boar struts off victorious. We slink out of the competition arena, heads down, but still alive.

As we walk towards the market, Antonio puts his hand on my shoulder. 'I believe you're the only Canadian to ever have competed in this festival. Good job.'

One of the parents of our volleyball team is an avid cinghiale hunter. At dinner one evening, he shares stories about his hunts, including times when his prize dogs met their fate as they were torn apart. As dinner finishes, he presents me with an authentic cinghiale tusk from one of his trophy hunts. I had the tusk crafted into a choker necklace for Bonnie.

Runaway Team

My cowboy friend Gordie and I are in the process of breaking a team of Percheron horses to pull a large hayrack. I'm working at Pioneer Ranch Camp in Sundre, Alberta – a Christian youth camp. We have about eighty head of horses in our herd used for riding activities and two teams of heavy horses for pulling wagons for hayrides.

We've been working on our young team of Percherons for about a month. Gordie is one of my best friends – the best man at Margie's and my wedding. He's an authentic cowboy, ranching in the foothills of the Rockies. His horsemanship is second to none, and he's responsible for me being able to sit a saddle. I can remember the early days of learning to ride when I'd saddle up and go out on my own for hours. I felt if I could only learn to sit a saddle while trotting, I could call myself a rider. I'd trot non-stop until blood soaked my Wrangler jeans – my tailbone skin rubbed raw; my tanned saddle turned crimson as if someone had dumped ketchup across it. We've been cowboying together for years, and I head out to his ranch every chance I get.

Gordie grew up in China, where his parents served as Christian missionaries. He's the kind of guy you would choose to go to war with – he'd never let you down. No matter what the odds, how hard or dirty the task, he would show up. His small frame encases a huge heart – one of empathy, love and gratitude.

One of the best ways to teach teams – horses, sled dogs or athletes – is to train them with more experienced teammates. Those veterans lead by example and help teach the rookies. We'd been progressively teaching the important skills to accomplish the tasks of a serviceable draft horse team – pulling their weight in the traces, responding to the bit and verbal cues, following the lead of the broken horse, remaining calm amid noise and chaos.

We decide it's time to put them to work. I'm the designated teamster preparing to take twenty teenagers out on a hayride with our young team. I harness the team and hook them up to the hayrack. I'm pleased with how calm they stand, but will my team behave as expected for this entire first outing? Having trained equine beasts many times before, I know that a rodeo can occur at any moment. You bet I'm fearful as I pull up to the excited crowd of young people. I explain to all this is the very first outing for this green team.

'If we happen to have a wreck, I want you to get off the hayrack as quickly as possible.'

I'm so happy with the behaviour of my team with all this commotion behind them. The rack is full of exuberant and boisterous teenagers, but the team keeps pulling and listening to my commands. I'm feeling very proud of the job I've done with Gordie, breaking this young team. They're performing beautifully. We crest a hill in a wide-open field and start our downward descent. Suddenly one of the wooden neck yokes snaps! (A neck yoke is attached to the pole holding the wagon back when descending.)

I now have nothing to hold back this wagon full of twenty screaming teenagers. The rack slams into the rumps of my team! They spook, rear up and break into a full gallop. Luckily this team has learned to obey my commands.

'Whoa, whoaaa.'

As I holler, they obey and slide to a halt. *Whack!* With nothing to hold the rack back, we bash into the butts of the horses once again. I turn to all on the rack and scream.

'Clear the rack.'

As the frightened horses once again blast off at a full gallop, I turn around to see the kids flying off my rack. The field is scattered with bodies of joyful teenagers.

I'm the only one on the rack as we hurtle full speed across the field with my runaway team. Now that everyone else is safe, I'm quite enjoying the exhilaration of this chariot race. I fantasise, *Here I am, a fearless Roman gladiator racing his chariot at breakneck speed in Rome's Colosseum – crowds cheering.* What a rush!

As we approach the gate exiting onto a thoroughfare, fear strikes as I realise I cannot have a runaway out on a busy country road. I spot a double row of trees lining the field and make the decision to turn my charging team into that row of trees to avoid crashing through the gate out onto the vehicular roadway. At full speed, I pull on my right rein – the horses respond and swing to the right – my wagon fishtails wildly – the stand of trees suddenly looms. We collide with the trees and come to an abrupt standstill. My body slams against the front of the rack.

The hayrack is wedged between two trees. The horses have hit with such force that one of them has flipped up and is now facing back towards the rack. I'm very proud of the horses who responded to my commands during this wreck. It's certainly no fault of theirs we had a runaway. With the wagon running into them as they followed my commands, I'm surprised they still listened. The good news – the team responded admirably; there are no injuries; the campers have an exciting tale to share with their loved ones; and I got to play chariot racer. Just another unplanned adventure in my life.

Wedding Carriage Ride

'Brad, I'd love to surprise my new bride. Could you please help me?' Richard is making plans for a special wedding treat for his fiancée, Brenda.

'Do you think you and Gordie could bring a horse and buggy into Calgary on the day of our wedding? I'd love to take Brenda from the church ceremony to our reception in Western style.'

Gordie and I haul the cowboy transportation in from Pioneer Ranch on wedding day. We pull our horse trailer and carriage into Britannia Slopes Park, Calgary. The escarpment drops off to the Elbow River sharply, not the ideal location for a warm-up – but close to the church.

'Brad, we've got time before the ceremony. Let's unload Doc and let him stretch his legs.'

We unload. Doc stands as we groom his hide, mane, and tail; clean his hooves; slide his collar over his ears; place his tack across his back; adjust his girth strap; pull his tail through the crupper and over the breeching; insert his bit; secure his bridle equipped with blinders; lift the shafts alongside; and hook the traces to the carriage.

By now, a large crowd has gathered – onlookers excited about encountering an equine beast in the middle of this urban sprawl. Gordie jumps up into the teamster box, and Doc glides admirably across the green turf with our polished carriage in tow, much to the delight of our crowd. Gordie looks like a seasoned carriage driver competing in the Calgary Stampede as he pilots Doc in circles and figure-eights.

'Brad, why don't you take off Doc's bridle and tie him to the trailer with his halter before I dismount.'

I pull off Doc's bridle. Now, without his blinders, Doc can see behind him! We've made a disastrous mistake. Gordie's slight movement startles Doc. He leaps forward. The carriage with Gordie aboard follows. It's a terrifying sight for Doc, who usually cannot see anything behind when hitched up with blinders on.

'Brad, grab him!'

I reach up and encircle Doc's neck with my right arm. He leaps, kicks, bucks, sways violently to the left and right, trying to get rid of the scary object chasing him. It's impossible for me to slow the frightened steed. We gallop across the park at breakneck speed – through gardens with fresh flowers, airborne as if targeted by a grenade. I feel like a rag doll flailing alongside this uncontrollable brute. I'm not sure who's more afraid, Doc or me? I struggle to put my fear on the back burner as I endeavour to come up with a solution.

I struggle to get my left hand up to his ear. *If only I can grab his ear and chomp down on it with my teeth, Doc will stop.* I've done it many times in wild horse races in my rodeo days. But no luck. I can't get my mouth to his ear.

The horrified crowd is scattering. Women are shrieking. Children are wailing as moms pull them to safety. The park is suddenly vacant as bystanders seek refuge behind parked cars.

Doc is streaking towards the escarpment – a drop of about thirty metres down to the Elbow River. I've got to find a way to stop this charging brute! A few more metres, and all three of us will be plummeting over the bluff. I attempt to settle my brain. Earing Doc is impossible. *What strategy can I use now?* I take my right leg and wrap it around Doc's left front leg. He stumbles, and plunges to the turf. We all come to a dusty halt. I stand to peer over the precipice, just two metres ahead.

'Great job, Brad. I thought we were done for.'

We unhook Doc, and I cool him off while Gordie checks out the condition of the carriage and harness. We still have a bride and groom to deliver.

We pull up in front of the church and anxiously await the arrival of the newlyweds. I'm making sure the blinders are on Doc. Gordie's driving. I stand in front of Doc. We don't need any more wrecks.

Scenes of the start of the Stampede chuckwagon races flash through my brain. Brenda squeals in delight. Richard sports a proud smile, and we trot down Elbow Drive towards the Glencoe Club reception as a procession of slow autos follows.

As we descend into the Elbow River valley, a piece of the damaged harness gives way. Doc looks as if he's ready for an encore rodeo.

'Brad, get out and grab Doc!'

I leap out of the box, sprint to the front, and grab Doc's bridle.

'Richard, get Brenda out of the carriage.'

The reception's great fun, especially since the gossip centres around the two cowboys who tamed a runaway steed.

Release the Beast

It all started a Pioneer Ranch Camp. As the camp waterfront director, I often head up to the corrals and hang out with the wranglers. Having been brought up in eastern Canada, it had always been a dream of mine to be a cowboy. Rodeo riders seem to be the epitome of a 'man's man' – an image I'd like to slip in to. On this evening, one of the wranglers suggests I climb into the bucketing chute and come out on a bareback bronc.

'Are you kiddin'? I know so little about horses. How'd I ride a bucking bronc?' I question.

'Ya don't have to know much about horses. Ya just gotta be a good athlete with great balance.'

Well, here's my chance to fulfil my dream of living the cowboy life. Straddling a hay bale, going through the actions of a bronc rider with instructions from my cowhand friends, I'm ready to tackle a snorting beast.

My friends load a bronc into the bucking chute. I stand behind the chute, observing this equine athlete. I wonder if I really want to follow through with this. *What's a waterfront director doing at the bucking chutes? What do I know about riding angry horses whose goal is to dump me?* I'm scared stiff!

My cowboy friends prepare me for the pandemonium about to occur. They set the bareback rigging, which resembles a suitcase handle attached to leather, on the withers of the steed. I load my leather glove with sticky resin and pull it onto my left hand. Wow, I don't even know if I ride left-handed or right-handed. I perch on top of the chute, looking down at my bronc.

I slide down into the chute with my legs straddling the 600-kilogram animal. I reconsider my request as my friend clarifies that waiting for the chute gate to open is dangerous – I could seriously injure my knees and ankles with a 'chute fighter'. I feel more secure as a cowboy grasps my belt to ensure I don't slip off this rearing horse, dropping me into the chute under my steed's hooves. The chute gate has not even opened yet, and this mount does not seem thrilled to have me perched on his back. My fear elevates as he rears back and strikes out with his front hooves. My cowboy friend clamps onto my belt. I exhale as I dangle from my belt, wedgy and all. The bronc settles down. I push and pull my rosined glove into my bareback rigging, set my spurs into the shoulders of the bronc, lean back onto my back pockets, say a prayer (I'm not sure if God talks to animals), tighten my grip on the rigging and nod my head. The gate swings open. There's no turning back from the chaos about to occur.

My steed explodes out of the chute, ducks to the left and rears back. I fight to maintain my balance and keep my upper body square to the horse. My spurring action is erratic as I try to emulate what I've practiced on the hay bale. My bronc gets some airtime, comes down and powerfully spins to the right. I get out o' shape. Both legs on the left side of his neck. He throws all that power into my ridin' arm, and now it's me who's gettin' airtime! I fly airborne and crumble into the dirt like a folding lawn chair – not even close to making the eight seconds.

Picking myself up from the rodeo arena dirt, I know I should feel defeated. The taste of dirt lingers in my mouth. My back aches. My left arm ligaments seem stretched. My tailbone feels as if it's been struck by a truck. I pick my Stetson out of the dirt and brush it off. Strangely, I have an exuberant feeling of victory. I'm brave enough to come out of the chute on this bronc and stay aboard for about three seconds.

Again and again, I burst from the bucking chute on top of one rank bronc after another as I try to improve my skills, reacting to his wild gyrations. I learn that every bucking horse is extremely intelligent and learns bucking tendencies he'll use again and again to rid him of his rider. He'll 'set the trap', and then it's 'ground time'. Once I'm out of shape, I ain't gettin' back into a position that will last the eight seconds. There's a lot of power coming at ya with every jump. I've got to get in time with the bucking, leaping, kicking monster. Finally, to the cheers of my rodeo friends, I cover a bronc for eight seconds. Bareback bronc is my new-found skirmish. I love the adrenaline rush before and after they swing the bucking chute gate open. Every time I take on the challenge, I say a prayer – help me handle my fear, be braver, be bolder, be better, God keep me safe.

***

It's the beginning of a new experience as I spend the next few summers riding bareback broncs in rodeos throughout Alberta. Perhaps the most prestigious rodeo I ride in is called Rodeo Royal – the Calgary Stampede's indoor rodeo in the spring. My draw calls for a Friday night ride.

As I'm ready to leave our ranch in Millarville, my wife Margie, nine months pregnant, states she feels 'tonight is the night'. I drop her off in Calgary with her parents as I proceed to the rodeo.

I finish my ride and call Margie's parents.

'How's Margie?'

'She's gone to the hospital.'

'Has she delivered yet?'

'We're not sure. You'd better get your cowboy butt over there.'

Yikes, don't tell me I've missed the birth of our fourth child! I feel so badly – I've put my own desires ahead of my family's once again!

I chuck my bareback rigging into the backseat, jump into the driver's seat, and barrel to the hospital. I park illegally, dash into the maternity ward in my cowboy boots and spurs only to witness Margie's disgusted frown. Too late. The baby's been delivered.

'Margie, I'm so sorry. What do we have?'

'It's a boy. Now we have two and two.'

Sometimes Teamwork is Essential

Our summer camp has just wound up, and in a few days, Gordie and I will trail eighty-five head of horses ninety-three kilometres from Rocky Mountain House to Sundre, Alberta. Our shortest route is to cross the bridge over the North Saskatchewan River, trail through Rocky Mountain House, then follow secondary roads through numerous small towns south to Sundre. We're not sure if the horses will cross the noisy, metal bridge into Rocky, so we're scouting the river on horseback for a shallow fording to use instead of the bridge.

Gordie is mounted on a very young horse, and I'm on a well-trained mount. We ride the west bank of the North Saskatchewan, looking for a suitable crossing. We discover a fording that seems shallow enough to cross. Gordie leads as I follow closely behind. As we move out into the river, the bottom drops away and our horses are forced to swim. Gordie's young horse hasn't experienced deep water before. He's never swum. His horse jumps off the bottom of the river, catches a breath before submerging underwater. After a few jumps, his horse's ears fill with water. He loses all sense of balance. He rolls over lifeless. Gordie, a non-swimmer, is floating downstream, clinging to his lifeless horse.

I feel helpless as my horse swims the width of the river. I'm not sure how long Gordie's horse will remain floating. I calculate my next move. I reach the opposite riverbank, and I start chasing Gordie downstream at full gallop. Gordie's floating down the middle of the river – beyond my reach. My only option to make contact is with my lariat. I unfasten my rope, closing the gap with Gordie. Fear fills me as I swing my loop over my head. I realise I'm not a great roper, but I just need one good throw. Gordie's eyes are fixed on me. I can see his look of panic. He's my best friend, and I've got to save him. I've got to make a good throw.

'Gordie, grab this.'

My first throw reaches Gordie. I've made contact. He slides the loop over the horn of his saddle. I hastily make a couple of dallies on my own saddle horn, anxiously waiting for my rope to become taunt. *Will my rope hold? Will my horse be able to withstand the force of Gordie and his horse?* I brace for the impact.

We successfully pull Gordie to shore, still grasping his lifeless horse. Gordie smiles and gives me a high five. His horse gives his head a good shake, sneezes and clears the water from his ears and nostrils. Teamwork with horse and rider got 'er done.

Our scouting excursion convinces us the bridge may be our best bet.

Killer Whales

My oldest son Bryn and I have joined up in Smithers, British Columbia and travelled down the Skeena River to Prince Rupert on the Pacific Coast. Travelling south through the Inside Passage by BC Ferry, we debark at Port Hardy, the northern-most port on Vancouver Island.

Bryn and I head south to the tiny village of Telegraph Cove on the eastern coast of Vancouver Island. We rent a two-person kayak, buy supplies, load up the kayak and paddle out of the protected bay. Our plan is to paddle south to the Robson Bight Ecological Reserve, where up to 200 orcas arrive each summer to rub on the barnacle-encrusted rocks at the mouth of the Tsitika River. The area is a popular whale-watching spot, and we hope our luck will afford us the sighting of resident and transient pods of 'killer' whales.

Not five minutes out into the strait, a cow and calf break the surface only three metres away from our kayak. What a way to start our trip! We paddle south, encountering five more pods of orcas in the next three hours. We pull onto shore and set up camp on a peninsula jutting out into the strait. The perfect site to scout north and south to see if whales are approaching.

Our routine for the next couple of days is to relax on shore until we spot a pod of whales passing by, then quickly jump into our kayak and paddle offshore, waiting for the whales to come. The pods seem intrigued by our presence and often approach in a curious manner, only metres away.

One afternoon, a pod of five whales stops ten metres off our bow. One whale rises vertically, poking its head out of the water, reminding me of a periscope on a passing submarine, has a look around and then slowly slides back down. It's not a breach, but a spy hop. Researchers believe it's a manoeuvre whereby they survey their surroundings for prey from their elevated position.

'Bryn, make sure you don't make any noise like a seal.'

As we return to shore, our neighbour campers come running over. 'You're so lucky to be alive. Those are "killer whales", you know.'

***

I roll over in my sleeping bag, awakened by an uncommon sound. Our tent is five metres from the upper tideline. I hear what sounds like continuous whale blowing. As I crawl out of our tent under a full moon, I witness an astonishing spectacle. A pod of six whales lying dormant ten metres offshore, blowing into the night air, a full moon casting shadows, water glistening under the bright moonlight and the blow of the whales reflecting the moon's light. I've never seen such a magnificent sight. The whales are sleeping on the surface and breathing rhythmically as they lie there. I rush back to the tent and wake up Bryn to share this experience. We sit on the shore, mesmerised. I admit it's possibly one time I feel a mystical connection to these remarkable marine mammals.

In our three days of paddling, we encounter more than 100 whales. Some of those sightings were repeats, but what an exciting trip.

REFLECTIONS. Magical Moments with Animals

Our magical moments with animals can range from our pets to animals we work with to wild animals. Not only are they exciting, but those animal moments can be precarious. We can never be sure how a wild animal will respond. The wilderness is their environment, their home, and they live by their rules. If I venture into their domain, I too must live by *their* rules.

I know animals, just like humans, are motivated by positive verbal feedback, but also by painful consequences resulting from misbehaving. When training horses, they respond to feeling – the bit in the mouth, leg cues, a hackamore over their nose, a whip, a pat on the neck.

I believe that training animals, coaching athletes and teaching children must follow similar steps. It's important to break down skill sets into progressive phases resulting in high performance. We must systematically teach skills from the fundamental basics to the more difficult. Our sequential teaching allows our followers to build confidence as they progress. In my experience, if those I lead have not had a chance to build skills with increasing degrees of difficulty, they will form bad habits to survive. The solidification of fundamental skills gives humans and animals the confidence to step into the next level with their new-found skills.

Once the technical skills have been mastered, we can start developing the psychological skills – skills that must be trained just like technical skills. I love to engage those I lead in psychological zingers, developing their mental toughness. These psychological challenges force followers to learn what technical tools they possess to overcome the barriers I place before them. I never want the barrier to be so great that they quit, whether human or animal, but rather train them to build their technical and psychological skill sets sequentially.

***

I had a horse who would rear up when showing me he did not want to work as directed, either through laziness or simply trying to show he was in control. I gathered my special coaching equipment – a stout piece of wood one metre long and a thin plastic bag full of warm water. As my horse reared up, I struck him between the ears with my wood, followed quickly by breaking the plastic bag of warm water over his head. The horse assumed that when he reared up, his head collided with something solid which caused him to bleed profusely as the water streamed down over his head. Lesson learned. This shocking consequence was something he did not enjoy, and he refrained from rearing.

***

As an international polo player, I was able to train my polo ponies to 'set me up on the ball' – to place me in a position to make the perfect shot. At full gallop, my horse would fight off an opponent as I focused on making my shot. It's the embodiment of two athletes working together to ensure success.

Italian Cinghiale **Bonnie's Cinghiale Necklace**

11. Nature's Power

'The mountains are calling and I must go.' **John Muir**

Kayaking the Sea of Cortez

I'm on an expedition with my son Bryn and his family – wife Christine, son Marcus (age nine), and daughter India (age eight). This is no leisurely paddle, but rather a challenging expedition as we circumnavigate the crown jewel of the Sea of Cortez, Isla Carmen – the largest island in Loreto Bay National Marine Park. Our outfitter explains the dangers of the trip and urges us to hire a guide. However, Bryn and Kristine are extremely experienced, adventurous souls, and the three of us decide to go it on our own.

The 130-kilometre circumnavigation around this isolated and uninhabited Mexican island is most dangerous as we traverse the north shores. Winds in the Sea of Cortez can pick up extremely quickly, forming two-metre swells as we're fully exposed to the rocky cliffs of the northern shoreline. Not only are we aware of the high winds and rough seas, but travelling on our own by compass and map necessitates us finding safe landing beaches for lunch and campsites – not an easy achievement.

We most often find a river valley etched into the high rocky shores, allowing us to pull up onto the pebble beaches to pitch our tents legally. The island is one of the few privately owned Mexican islands. The rumours indicate the island is owned by Carlos Salinas de Gortari, Mexico's ex-president and one of the richest men in the world.

Our outfitter transports us and our gear to the beach of Puerto Escondido, south of Laredo. She looks on in disbelief as we load our mountains of supplies. Perhaps the heaviest cargo we load is drinking water. Since there is no fresh water on the island, we must carry five litres/paddler/day for the eight days. Wow, I had never envisioned carrying the weight of 200 litres of drinking water. Add to that our snorkelling gear, cooking utensils, gas stove and fuel, sleeping bags, air mattresses, tents and food (including our Coronas and limes). You can imagine how quickly the sheer volume and weight of our cargo fills our kayak storage hatches.

We paddle two double kayaks and a single. Bryn and I each have one of the kids as our paddling partner, while Christine paddles solo. We struggle to hoist our overloaded kayaks into the water, slide into the cockpits and push offshore. Although I've had extensive paddling experience, I'm so happy to have Bryn and Christine accompanying me on this perilous trip. Both are extremely experienced outdoor experts. Bryn spends six months every year in the outdoors as the CEO of the largest silviculture company in British Columbia. Christine is a former student of mine, having graduated from the Outdoor Pursuit Program at the University of Calgary. Their experience in wilderness travel and survival is certainly a comfort.

We take our first paddle strokes mid-afternoon and head east to Isla Danzante, ten kilometres offshore. The waters are crystal clear as we gaze below the surface along the shoreline at the abundance of tropical fish and stunning coral. It's as if we've stepped onto another planet. An unknown world full of sea creatures and monsters, deserted mountain landscapes and very few sand beaches. My body trembles with excitement and trepidation as I paddle into the uncomfortable, stretching my adventurous spirit to explore this new world. It's an odd feeling being out of my element surrounded by creatures that are most definitely in theirs.

I giggle a little as I paddle into our campsites. We are so spoiled coming from a country dotted with pristine, manicured campgrounds, equipped with running water and hot showers. As we pull up onto shore for the night at Bahia Hoonda, our routine becomes second nature. Every afternoon before we get to work on our campsite, Christine breaks out the Coronas and an exotic cheese snack. Christine is a wilderness cook second to none, having cooked in bush camps with crews of 300 labourers. We are spoiled to have her expertise on our trip.

Bryn and I set up the three tents, Christine sets up the kitchen and the kids blow up our air mattresses. We each throw our dry bags and sleeping bags into our tents. It's now time to explore our surroundings or jump into the water for a swim or snorkel before coming back for dinner. I can't express how refreshing it feels to plunge into the saltwater, eager to cool off after a full day of paddling in thirty-plus-degree weather, fully exposed in our kayaks. As I splash into the water, I feel instant refreshment as the cool waters caresses my skin. Entering the water, I do the 'stingray shuffle' – sliding my feet through the sand bottom, watching the stingrays as they scurry away. Stingrays sting with the sharp barb found in their tails that carries a protein-based venom. Venom enters through the wound, causing short-term, but intense discomfort.

Falling into bed after a scrumptious meal feels so good. Even at night, the heat does not dissipate as I lie in my tent – *Do I get inside my sleeping bag, or do I lay on top exposed to all those creepy creatures?* Isla Carmen is known for its abundance of scorpions and tarantulas.

As I awake on day two, I find I've aching muscles where I didn't even know muscles existed. Another day of adventure awaits us. We break camp and load our kayaks. The breeze is gentle, and the sun kisses our tanned faces.

The island is stunningly beautiful – cliffs of blood-red basalt rising sharply out of the sea and the bold outlines of towering cacti providing the backdrop for our paddling. Rounding Punta Baja on the southern tip of the island, we head up the east coast, arriving at our campsite at Arroyo Blanco – a long day of paddling sixteen kilometres.

We're not alone in our water environment – sea lions, tropical fish, whales, dolphins, manta rays, pelicans, blue-footed boobies, ospreys, frigates, herons, and egrets. It's prohibited to hike on the island, unless, of course, you have paid tens of thousands of dollars to hunt bighorn sheep – another illegal income for ex-president Salinas de Gortari.

'Look out to starboard side. What's all that splashing?' I shout.

'Wow. It's a pod of about 400 dolphins leaping across our bow,' responds Marcus.

We sit in awe as the pod passes, their 1.9-metre streamlined bodies advancing at twenty kilometres/hour. They fly through the air – engaging in spectacular aerial jumps, splashing down again, diving under the surface only to explode upwards once more. We follow the dolphins into the bay to observe them feed on a school of triggerfish, coming up close enough for us to pet. A juvenile nibbles my arm in a playful way and swims circles around our kayak. We listen to the whistles and clicks as they communicate with one another.

Midway around the island, we'll make a stop at Salinas Bay to visit an abandoned nineteenth-century salt mine and coral chapel. What an epic expedition. The ocean is not one of my comfortable environments – the swells, wind, tides, currents, sea life and compass work. It's all a challenge to me. What a blessing to share this journey with my family. Embarking on dangerous expeditions with family increases the fear level for me. It's not that every person I lead is not valuable, but being responsible for the safety of family reaches a new level of concern. I've experienced the horrendous grief that lingers with the loss of two sons too early. Once again, I learn the value of travelling with experienced partners – Bryn and Christine. It's an expedition chiselled into my lifelong memory. I check off another bucket list adventure.

Pacific Storm

As an outdoor pursuit instructor and guide, I sometimes find it difficult to understand the fear embedded within my students. I decide it's time for me to step out of my comfort zone and experience an outdoor adventure that will put fear into my heart. I feel I would be a more empathetic leader if I could understand the fear my students are facing. I decide to take my wife Bonnie, seven-year-old son Brett, and five-year-old son Justin on a week-long yacht trip in the Pacific Ocean.

The west coast of British Columbia is a magical place if you have a capable vessel, an experienced captain and crew. We pick up our twelve-metre motor yacht in Vancouver, and head north 200 kilometres to Desolation Sound.

As captain of our motor yacht, I must be ready to understand marine map reading, VHS radio operations, depth charts, marine rules, tide charts, tide currents, setting anchor, docking, and the mechanics of our engine. To be honest, I have very little experience in all these areas. Yes, I have captained a barge 118 kilometres through forty-five locks on the Midi Canal in France, but very little ocean experience! We receive a five-minute orientation session on the yacht docks and set off on our own down the Fraser River. My heart pounds swiftly as we leave the mouth of the river and enter the wide expanse of the Pacific. I face one of the most fearful and daunting tasks of my life.

As we exit the Fraser, we're immediately challenged with navigation. Using map and compass, we must chart a safe course as we sail to our next desired destination. Each evening, Bonnie and I pour over our marine charts and plan the next day's excursion. It looks so simple on a map, but we must depend completely on our compass and map-reading skills. Occasionally, we witness a small town or lighthouse with a distinguishing flash. Otherwise, we attempt to pinpoint our location with our compass.

'Brad, it's getting dangerously shallow!' Bonnie calls. 'Our depth finder is showing only twenty metres. No, now it's only ten metres. Stop!'

According to our navigation, we should be in water that's at least sixty metres deep at this point. Running aground would be disastrous. We struggle to pinpoint our exact location as we try to decipher where we are in relation to the shoreline. Not an easy task when everything looks the same. I'm terrified! Progressing very slowly with Bonnie monitoring our depth finder, we eventually sort out where we are – mission accomplished.

At the end of the day, as we pull into a harbour to dock for the night, it seems every boater comes up on deck to watch us dock. It's such an intimidating experience – trying to dock in a narrow opening with all these experienced boaters scrutinising! I'm on the flying bridge, controlling the motor and rudder while Bonnie is down on the deck looking after our bumpers, ready to jump ashore with rope in hand. Thankfully, I've had the experience of docking our barge in France dozens of times, but here the wind plays havoc with my game plan. More than once, I frantically call down to Bonnie to change the bumpers to the opposite side as the wind swings us bow for stern. I feel relieved as we tie up to the dock for dinner and a well-earned sleep.

Justin and Brett have heard the numerous stories of the gigantic salmon caught off the coast of BC. We pull out the tack, bait the hooks and drop our lines into the water as we troll for a prized BC salmon.

'Dad, I've got one!' Brett exclaims.

All hands-on deck. I cut the motor, Justin grabs the net, and Bonnie pulls out our *Fishing the Pacific* book as Brett reels in his catch. The fight in the fish is perfect for our seven-year-old son. As his five-minute tussle nears an end, we prepare to land his first saltwater trophy. As we peer down into the crystal-clear water, Brett's catch appears ominous. His dream of landing a huge salmon seems to disappear. It seems he's caught a shark! Bonnie quickly references 'how to land a shark' while ordering Justin onto the safety of the upper deck with her. Brett successfully tugs the metre-long fish onto the lower deck of our boat. Bonnie quotes the warning from our fishing encyclopedia.

'Sharks are extremely dangerous. Do not bring a shark on board. Cut it loose.'

Brett is not about to throw his first ocean catch overboard. He attempts to secure the fish and pull the hook out of its mouth with my help. Subduing a thrashing shark is no easy job. Justin's eyes grow bigger as our catch flips and flops dangerously close to our legs.

'Throw it back. Throw it back.'

After many attempts, Brett finally connects with a swift blow to the head of the shark. It lies unconscious on the deck. Game plan accomplished – our first fish landed. On closer examination and referencing our fish guide, we realise Brett's catch is not a salmon nor a shark, but a dogfish. Proudly holding his first catch provides us with our first photo op.

After a couple of days on the ocean, Bonnie and I decide to take the boys for a shore excursion. The four of us jump into our small tender and row ashore. As we scour the beach for seashells, Bonnie's panicking voice blurts out, 'Our yacht's broken anchor – it's drifting towards those other vessels.'

'Everybody – into the tender!' I command.

We push off the stony beach and start rowing frantically towards our yacht.

'Row faster, Daddy. Our yacht's going to crash into another boat!'

I'm pulling as hard and as fast as I can on our stubby oars. Bonnie's face expresses the fear she is feeling. I'm not sure if the boys are excited or terrified. The gap between our boat and the anchored boats is closing quickly. It seems hopeless. I don't think we'll reach our yacht before it collides.

'Do you realise how much it will cost if our boat crashes into that other boat? This trip is getting expensive!' claims Bonnie.

As we prepare to listen to the crunch of our stern into the deck of the anchored vessel, the engine on our boat starts and it pulls away, out of danger. Can you believe it? Another sailor witnesses our boat break free and quickly boards, enters the cabin and starts our engine. Another boating mistake by Brad – but this time, I'm lucky. I'd left the key in the ignition, and we're very grateful for an alert yachtsman.

I find entering Desolation Sound a stressful passage, navigating between the reefs and the shoreline, watching swells break over boat-eating rocks to port and starboard. We feel very proud of ourselves for having reached this destination while still afloat and undamaged. We've learned many marine lessons, but still have much more to encounter and learn. I'm so grateful for Bonnie and the boys, who've been so supportive. The fjord-like setting of Desolation Sound is everything they say it is – crystal-clear water, incredible mountain landscapes, isolated sand beaches and private coves tucked beneath the snowy mountains. We all go for a swim in a spot that is a freak of nature – the warmest saltwater north of Mexico. This far north, we plunge into water that is 24C.

Our trip is half over, and we head south towards Vancouver through the Strait of Georgia. Our days have been filled with fishing, swimming, beachcombing, walking through rainforests, exploring inlets, lagoons, estuaries, sea caves and sea stacks. It's our last day on the water before returning our boat on time into Vancouver harbour. Our course takes us east of Texada Island, through Malaspina Strait – a piece of water known for its violent storms. Channel 16 on our VHF radio blurts out the news, 'Storm brewing off Texada Island with high-velocity winds and seas building 2-3-metres.'

My brain says I should stay moored in the safety of the harbour, but we must return the boat on schedule. As we venture out into the channel, I can feel the howling wind and see nature's wildest weather unfolding on the ocean's surface. We're heading straight into the gale and crashing through massive swells, banging down on every wave. Brett and I are high up on the flying bridge, controlling the boat, with Bonnie and Justin seeking refuge in the cabin below. Our yacht lurches and rolls in the storm. I hear dishes crashing to the floor as they catapult from the cupboards and open shelves. Bonnie and Justin are being slammed against the cabin walls. Bonnie opens the cabin door and calls up, 'Brad, there are no other boats out here on the water. What are *we* doing out here?'

The storm continues to churn. I'm struggling to maintain control of our boat, with one hand on the gas and the other on the wheel. Brett is helping by holding the map open on the consul so I can tell where we are. I now fully understand the fear my students encounter when they find themselves in an uncomfortable situation – I'm frightened. The waves are breaking over our 2.5-metre bow. Water is flooding over the gunwales. The wind is whipping the salt water into our faces, stinging our eyes.

As I have so many times explained to my athletes and students – 'It's not the cards we are dealt, but how we deal with those cards.' When fear sets in, I must not become crippled, but instead return to the fundamentals of what I'm trying to accomplish. I must stay focused and not let my mind wander. *How much can this boat handle? If I turn sideways into the trough of the waves, will we roll? Will this boat swamp in these high seas? What are our chances of being rescued if our boat sinks?* Instead of thinking of these catastrophes, I must focus on my task of maintaining control of myself and the boat.

Brett and I discover an alternative leeward passage inside Nelson Island off the port side of the boat, allowing us to travel in sheltered waters.

'Brett, I can't turn sideways and expose the starboard gunwale of our boat to these high waves. We could roll! Hang on, and we'll try a 180-degree turn on the top of this next wave and run downwind in behind Nelson Island.'

Turn we do as we violently heel over to portside. We've successfully made the manoeuvre and ride the waves into the calm waters behind the island.

We pull back into Vancouver and tie up to home port. I've plunged into that state of fear which each one of us encounters when we feel we're no longer in control. It's a great lesson for me – to feel swallowed up by fear, but then be able to focus on what I need to do in that moment to find success – 'control the controllable'. I feel better able to return to instructing and leading with a new sense of empathy for those I lead as they face their fears.

Caught in a Hurricane

'Raft up! Now! Paddle hard! That hurricane's coming!' I bark.

The boys fight the wind and struggle to come together.

Rafting up is a manoeuvre accomplished as the group of canoes comes side-by-side, grasping each other's gunwales. The effect is a stable platform across the canoes with very little chance of tipping.

'Lay down in the bottom of your canoe.'

The wind howls over our canoes as our centre of gravity is significantly lowered. The whistling of the wind reminds me of a news bulletin during a Caribbean hurricane. The white-capped waves crash into our gunwales. The spray soaks our faces. I hear the boys whispering. I'm praying.

'Hang on tight. Don't let go of the gunwale. We're safe as long as we stay low and hold onto that next canoe.'

'Did you see that!'

'Look at that tree!'

'Hey, that car just got smashed!'

People on shore are screaming. Crashing trees splinter as they smash everything in their path. The roof of a shed is ripped off and flies into the trees. We hear breaking glass as windows are blown in. I'm not sure if the boys are excited or terrified? I'm scared.

I'm teaching canoeing on Crimson Lake near Rocky Mountain House, Alberta – eleven boys and myself in six canoes. I'm putting my years of canoeing experience to use by volunteering my summer as the Waterfront Director at Pioneer Ranch Camp. My days are filled with activities on the water – swimming, canoeing, sailing, waterskiing, windsurfing and life-saving certification. My swimming and canoeing programmes are composed of two levels – swimming certification leading to life-saving skills, and lake canoeing leading to white-water paddling. The capstone activity with my canoe programme is a five-day white-water trip down the North Saskatchewan River from Rocky Mountain House to Edmonton.

My own canoeing adventures started before my young sister, younger brother and I were teenagers. Dad was a teacher, so we spent most of our summers enjoying adventures with my parents either at our cottage or car camping. As soon as my younger sister could walk, the five of us would head out on the water. We canoed Northern Ontario on many exciting ventures. I still remember my father saying, 'I can hardly wait until you boys can carry the canoes on our portages.'

Paddling with my parents entrenched a skill set enabling me to get summer positions as a canoe trip guide in youth camps. I would spend six days a week paddling the waterways of Ontario. Eventually, my skills enabled me to paddle on Canada's International White-water Canoe Team, become an internationally recognised white-water rescue expert, direct a white-water rescue film winning the gold medal at the Banff International Film Festival and land me a full-time position at the University of Calgary instructing in the Kinesiology Outdoor Pursuits Program. I never imagined those days paddling with my parents would take me on a path that would gain international recognition and a full-time university position.

Our canoes are pushed towards the shoreline, but before we touch land, the winds stop as quickly as they had started. I peer over my gunwales to count our flotilla, *1, 2, 3, 4, 5, 6*. All my canoes are here with me, upright and floating with the eleven boys laying beneath the thwarts. I thank God. My heart is still wildly beating as I tell the boys they can sit up. Their facial expressions have dramatically changed from fear to excitement.

'Hey, we rode out a hurricane! This may have been the safest place to be in that hurricane!'

'Wow, we did it. I can hardly wait to tell my parents about this wild ride!'

After the excitement subsides, I debrief the adventure with the boys by asking them about our safety procedures and how they controlled their fears as they obeyed my commands.

'Brad, I knew we were safe with you.'

We paddle back to camp to survey the damage, hoping none of our fellow campers have been injured. I congratulate the boys on their stellar performance as we arrive back in camp safely.

I'm pleased with the progression of the paddling skills displayed by these twelve-year-old boys out on the lake today. In an instant, everything dramatically changed. A category-1 hurricane, with wind gusts over 100 kilometres/hour sweeping across the lake. From our position in the middle of the lake, we could see terrified vacationers fleeing from their cabins. It felt like we were in a movie theatre, watching a horror movie!

I'm a firm believer we must *not* protect our youth from risk, but instead train them to manage risk. We must also understand that taking calculated risks necessitates coping with fear. If the risk is extremely dangerous, we may not be equipped to handle that fear or danger. In that case, it's best to avoid the risk. Sometimes, it may not be possible to step away from danger. It's crucial we have developed the skill set to handle those challenges.

The Ram River

My students are in their graduating year and have completed two university courses – lake and river canoeing. To graduate, they must successfully complete this capstone activity requiring them to exhibit skill sets in canoeing, climbing, camping and survival. The expedition will test our students to the fullest. Completing the expedition will not be easy.

The Ram is not a big volume river, but a very technical river to paddle – a five-metre-wide white-water river with continuous class 3–5 rapids, three major sets of waterfalls we must rappel down, a 550-metre river canyon, rock cliffs with treacherous undercuts, sweepers, and log jams, and harrowing portages. Every corner is an adventure waiting to be explored. Not for the faint in heart. Even more fearful for me as a leader. My previous hours of prep with these students give me confidence.

We descend the shale slope and drop our canoes into the water on river left, load them up, and prepare to ferry glide to the opposite shore. Expedition partner Bill March (the leader of Canada's first ascent of Mt. Everest) and I will ferry glide across the raging Ram River above the Ram Falls (a thirty-metre waterfall) and set up our safety positions. It's essential every boat traverse the current safely before plunging over the deadly drop. The roar of the falls reverberates in my ears – similar to a thundering freight train. The plume of mist rises ominously above the lip of the cascade. We ease out of the calm eddy into the rapid current. The river seizes the bow of our canoe and tries to rotate us downstream as we fight to keep our upstream ferry glide angle. Our strokes accelerate against the current. Our angle is precise. We glide safely into the eddy on river right and jump onto shore. One boat across, six more to go.

Twelve students in tandem canoes must traverse the current into our eddy. Bill and I set up our safety positions on the edge of the eddy, throw bags in hand just seven metres above the falls. The canoes will come one at a time. It's our first major challenge. The students are nervous. Bill's anxious. I'm afraid. This first manoeuvre is immensely stressful and will be a good indicator of the success or failure on our four-day, ninety-five-kilometre descent of the Ram River. If one of the canoes tips or misses our eddy, Bill and I have just one throw with our throw bags to rescue the students before they are swept over the thirty-metre drop! As I stand poised on the riverbank, throw bag in hand, fear grips me as I comprehend the enormity of this dangerous endeavour.

'If a canoe tips, I'll throw to the first victim, and you throw to the second Bill. If either of us miss our throw, we're taking the student home in a body bag!'

The first three canoes pull into our eddy successfully, and we signal for the next canoe to commence. The fourth canoe inadvertently broadsides on a large rock in the middle of the river twelve metres above the waterfalls. The students are stranded on the rock. The canoe is pinned. Bill and I sprint upstream. We throw our throw bag to one of the students and pendulum her to the safety of our shore. The other student assists us as we set up a mechanical advantage rope rescue of the canoe and stranded victim. What a relief as the second student steps ashore to cheers and high fives. The fear invading my body will repeat itself again and again on this trip as I accept the overwhelming leadership responsibility.

We prepare to set up a static zipline and rappel station to descend the waterfalls. With our rappel anchor prepared, a couple of students rappel to the riverbed below to set up an anchor for our zipline. Our loaded canoes slide down the zipline, and we rappel. This first trial is exactly the desired purpose of our capstone expedition – to challenge our students fully.

It seems as if we've just started paddling when we come upon Tapestry Falls, a twenty-five-metre drop. Our zipline and rappel go smoothly as our students become more familiar with the procedure. We paddle a class-3 section of rapids before pulling out. A perfect spot to camp in this scenic canyon after our first day of overcoming the challenges of this wild river.

I awaken on our second day, pull back my frost-encrusted tent flap and poke my head out to greet fifteen centimetres of snow. Just what we need to add to our struggles. The morning sun is up, peeking over the canyon walls, bathing the river gorge with golden light and warmth. I crawl across the mouth of my tent, my hands freezing as they contact the snow-covered moss.

We get the fire going and warm our frigid hands. I can feel hot coffee trickle down my throat and warm my belly as we prepare breakfast. We devour breakfast, break camp and load our boats. Our wetsuits are frozen stiff, standing upright in the snow like scarecrows guarding a farmer's crop. We pour hot water down the inside until they become pliable. I struggle to encase my body as I pull the tight neoprene suit over my torso. A student zips up the back. I kneel to slip on my wet boots, pull on my neoprene gloves, strap on my personal flotation device and I'm ready.

Day two starts with Table Rock Falls, a twelve-metre drop. As the first rays of light break through the snow-laden trees, the alpine flowers spring back to life. Once again, we use our expertise to lower our boats and rappel into the river below as we paddle towards the first of two canyons with 500-metre vertical walls rising from the river. Ricochet Rapid is class-5 – a death-trap. It's impossible to run. We must portage. The unmarked portage follows a narrow sheep trail angling across an exposed face of the cliff. Bill and I set a rope to hang onto with one hand while carrying our boats and gear with the other. One misstep would result in a slide down the steep slope and over the cliff into the deadly water. At the end of our portage, we rope up the bow and stern of our canoes and lower them four metres into the river. We then rappel down to position ourselves in our canoes as the current tries to sweep away our boats.

Bill is the 'tail' on our expedition in his kayak, and enters the current last. He is immediately swept under an undercut on the opposite side of the river. The current has him pinned under the overhanging rock cliff.

His boat is wedged under the rock. The river traps him in the grip of the undercut. Self-rescue from an undercut is nearly impossible. The students look on with panic. I realise my best friend could drown before my eyes! I grab my throw bag and make my toss to Bill. His fingers clench the bag. I dally my line onto my thwart. I'm pulled downstream as the current seizes my canoe, extracting Bill from the undercut. Bill rolls up and gives me a high five. We continue downriver.

It's our fourth day of paddling, and I'm proud of our students for managing to conquer the obstacles we've faced. They're thriving on the challenges we're facing. As the lead canoe, I pull into an eddy to wait for the group. Sitting in the quiet water, I notice a log covered with moss floating beside me. I push it underwater with my paddle. As it resurfaces, I realise it's not a log, but a dead black bear! *Wow, how can I utilise this bear in a teachable moment – a learning opportunity that will stay with my students forever?*

'Jump out and pull our canoe up onto the shore,' I shout to my student paddling partner.

As he is securing the canoe, I grab the bear and pull it onto the beach. The bear is about a three-year-old black, not very big. As I look at it, I'm not sure how it died. I don't know if it fell off the cliff or drowned in the raging river. I do realise it's a recent death since the bear is not bloated. Clutching the bear, an idea flashes through my brain. *Let's see who separates themselves from the group as they show critical and creative leadership thinking.*

To the astonishment of my paddling partner, I collapse onto my back and pull the bear on top of me. His surprised look makes me wonder if I should follow through with my plan. It's too late now. The first canoe rounds the corner and spots us on the shore.

'Help! Help! This bear's got me!' I cry.

I hear the first tandem paddlers, 'It's Kilb. Keep paddling.'

Needless to say, those two students failed the class (just kidding). Canoe after canoe pulls up. Most students don't leave their boat. I continue to flail on the beach under the bear. I hear their panicked statements.

'What should we do? Why doesn't that bear run away with all of us here? How do we separate Kilb from the bear? I'm not risking my life with that bear!'

A few students cautiously proceed toward me – their anxiety changes to laughter. They realise I'm up to my old tricks – grabbing a teachable moment as I try to discover leadership talents and relieve the tension of this intense trip.

Another idea pops into my head as we prepare to leave.

'Would anyone like a bear's claw as a keepsake of this trip?'

I extract two claws for my own boys and share the remaining with students. It's these precious moments that keep our dreams alive, our sanity in check, and our hopes fuelled.

As the leader of this expedition, I believe I must impart more than canoe strokes. I agree with the research that points out that the best education is 'experiential education' – education in which students are actively engaged in challenging activities leading to creative, critical thinking. Experiential education not only allows us to develop skill sets appropriate to various activities, but also experience fear. Setting up events introducing fear within the confines of a safe environment and the support of leaders can help to prepare our charges (or children) with progressive insights on how to 'dance with fear'. As a leader, guide and coach, I'm constantly looking for teachable moments where I can empower followers to act in a leadership role. Often, these teachable moments are not planned in my lesson plan. They may be moments that occur unpredictably throughout the course of the activity. I'm sure my coaching background has aided me in using these valuable learning opportunities.

I cannot express how relieved I am to pull into Rocky Mountain House with all students intact. Our outcomes include enhancing trust, cooperation, self-confidence, self-esteem, positive risk-taking, appropriate decision-making, leadership and teamwork. In this life-or-death scenario, true character is exposed.

This was our first and last trip with students on the Ram River. Both Bill and I believe that, although this epic expedition tested our students physically and mentally to the fullest, we were lucky to come off the river without incident. The Ram is just too dangerous. Yes, we were prepared. Bill and I, both expert paddlers, had run the Ram just days before our expedition for a thorough reccy. The students had worked hard to acquire the skills necessary to survive, but there are just too many factors out of our control. One mistake, and we could have lost a life! We're not willing to take that chance again. As leaders, we must carefully weigh the benefits and dangers. Often, the higher the risk, the greater the outcomes. However, there comes a point whereby fear and dangers are simply too grave, and we must say 'No.'

REFLECTIONS. Nature's Power

Nature is not always pretty – she's wild, violent, unpredictable, dangerous and unforgiving. Her rules are constant and absolute. They don't change for any of us, regardless of ability, experience or desire. One important lesson I've learned through my numerous experiences in the outdoors is the fact that fighting Mother Nature is a losing battle. Rarely will we win. She's so strong, we're simply over-matched – defeat is not 'if', but 'when'. Instead of battling, I've learned that sometimes I must harness her power to become successful.

As a white-water paddler, I've learned I'm best if I live like water – the way it bounces off obstacles and flows around rocks on its downstream journey. As I emerge from a quiet eddy into the torrent, I use the current to help turn my canoe downstream. I can use the current to move me from side to side with ferry glides. If caught in a hydraulic, I must give myself over to it to get flushed out along the bottom of the river. If surfing, instead of stroking through the breaking waves to get to the 'outside', I can catch a riptide which will whisk me along as if on a moving sidewalk in a busy airport. If I get caught in a riptide while swimming, I must go with the current as it carries me away from shore, swimming diagonally out to safety. When skiing, I use gravity and the fall line to assist me with my carving action while turning. In all these cases, I attempt to make use of the power of Mother Nature rather than fight it.

The challenge of making the right decisions in stressful situations can sometimes depend on not what we know, but *how we use what we know*. There's no question in my mind that fear plays an important role in making correct decisions. Fear is my inborn warning system – letting me know what I must pay attention to. The trick comes to handling that fear. Does fear take me out of my full-functioning mode, or can I use it as an additional tool helping me to be successful – accepting our fear as a reinforcing companion?

I find that every rescue scenario is different. It's important for me to develop principles I can utilise, rather than memorising step-by-step rules of what to do. It's important to be flexible and adapt rescue procedures based on the situation I find myself in.

One of my favourite formulas for preparing to deal with mishaps is an exercise I often implement when participating in an activity – I visualise a challenge and cognitively think about what I would do, *If a canoe tipped with two paddlers? If a paddler were pinned against that bridge abutment?* This rehearsal fills my toolbox with options and prepares me with swift and effective actions.

My reputation as an international white-water rescue expert was never a goal I'd ever dreamt of achieving. The recognition was very humbling and occurred by accident. It was a process of rescuing victims every weekend as I taught and led literally dozens of river classes and expeditions.

***

In 1983, the Canadian Life Saving Society commissioned me to produce the film, *Rescues for River Runners*. The film was an outcome of my whitewater leadership and rescue knowledge. The film won the gold medal (Best Film on Mountain Safety) at the *Banff International Mountain Film Festival*. Following the recognition of my film, I was invited to lead many white-water rescue seminars not only in Canada, but also internationally with search-and-rescue personnel. It was an outcome I'd never planned to pursue, providing a superb growth experience for me as I worked with numerous professional search-and-rescue teams in North America and Europe.

***

The Matterhorn soars above Zermatt near the Swiss-Italian border at 4,478 metres. It's perhaps the most photographed and recognisable mountain on the planet. Approximately 3,000 climbers undertake to summit every year, with an average of twelve climbers losing their lives annually. Mountain climbing sometimes becomes a spectator sport. When someone is stuck on the face, word quickly spreads and spectators head to Zermatt by droves to view the show. The element of danger makes good theatre. Many glass through binoculars while having lunch or a drink.

While on a ski vacation in Zermatt with my family, four foreign climbers came to attempt a winter summit before the winter season finished. The climbers attempted to hire a local guide, but no guide would accompany them due to the weather forecast. The climbers decided to continue with no guide.

'Guides turn back in bad weather; guideless parties continue. It's foolish,' claimed local guide Kronig.

On the second day of climbing, a full-force storm hit the mountain. I still remember joining large crowds of locals glassing the climbers as the clouds and snow eased. They were bivouacked on the face – no movement. On the third day – still no movement. The locals became restless.

'I trust no guide risks his life climbing up to save those fools. He would definitely die.'

Die they did – all four of them. Sometimes adventurers take undue risk without carefully paying heed to the factors that are out of their control – not a wise decision. After assessing the danger, I may have to park my ego at the trailhead or turn around and retreat. Those of us who are so goal-oriented may find this difficult, but those of us who make foolhardy decisions do not survive into old age.

Family Barge, Midi Canal, France

Family Yacht Excursion, Pacific

Brett's 'Shark' Catch

Pacific Yacht Captains, Justin, Brad, Brett

Sea Kayaking, Sea of Cortez, Mexico

Ram River Bear

12. Quick Thinking

'Be quick to take advantage of an advantage.'   H. JACKSON BROWN, JR.

Runaway Trailer

What an amazing fall afternoon on Glenmore Lake in Calgary. Golden sun blazing down upon us. Azure blue water dotted with whitecaps. Bright yellow autumn colours cloaking the lake. A perfect day for my University of Calgary canoeing class with fifteen students.

My class focuses on self-rescue, assisting rescues, bow and stern tandem strokes and solo strokes. Canoeing is one of my favourite outdoor activities. Mom and Dad started me canoeing at our cabin in Ontario, teaching me the fundamentals. At age ten, we started taking week-long summer canoe excursions in Northern Ontario. By age sixteen, I had enough experience to start guiding at youth camps. Getting paid as a teenage guide on the water six days a week is a pretty special way to spend my summers.

All those hours spent with a paddle in my hands lead to a career at the University of Calgary. Fifty percent of my job description involves coaching the varsity volleyball team and fifty percent instructing within the Outdoor Pursuits Program.

Today we've spent three hours on the lake paddling tandem. It's time to peel off our personal flotation devices, throw our paddles into the canoe trailer box and load the canoes onto the trailer. We pull our boats up onto the shore, but I must fetch my truck and empty trailer that's parked at the top of a long boat ramp.

'Hey, Brad. Can I go up and drive the truck and trailer down the ramp for us to load?' asks a student.

'Of course. Here's the keys.'

He drives down the ramp, then executes several back-and-forward manoeuvres to turn the vehicle around, ready to head up the ramp again.

I ensure the canoes are secured, do a quick sweep to make sure no equipment is left behind, and jump into the driver's seat. I crank the radio to listen to one of my favourite Western songs as I head back across Calgary to the university campus. As I descend into the Bow River valley along Crowchild Trail, I glance out my driver's window. *OMG, there's a canoe trailer passing me. I've got to get away from this runaway before a wreck happens.*

My instinct tells me there's going to be a crash with this loaded canoe trailer flying down the freeway. *What? That's my trailer!*

My grip on the steering wheel intensifies. I've got to do something to abort this disaster. *How do I stop this runaway trailer? This loaded trailer could kill someone as it slides along on its tongue!*

I hit the accelerator. Get in front of the trailer. Cut violently in front of it. Focus on my rear-view mirrors. Delicately slide the tongue of the trailer under my rear bumper. Gently apply my brakes. The trailer comes to rest wedged underneath my bumper. I did it. I engage my emergency lights and get out to survey the damage.

Unknown to me, as my student manoeuvred the truck and trailer at the bottom of the ramp, he jackknifed. As I hit a bump, the damaged hitch disengaged, the safety chain snapped and the trailer flew past.

I use my safety chains to secure the trailer and drive out of the river valley to campus. My most embarrassing moment is as I step out of my truck with large wet spots on my shorts! And I thought the most dangerous activity today would be canoeing on the Glenmore.

Caught with My Pants Down

I'm coaching at the National Volleyball Championships. I dash to the washroom for my nervous excretion before the match. I hustle in only to find all the stalls full, except for the handicap stall. With time at a premium,

I decide to use the handicap stall even though I'm an able-bodied coach. As I squat on my throne, I see the wheels of a wheelchair pull up in front of my stall!

Oh no, caught with my pants down! I feel a combination of fear and guilt as if I've parked my car in a handicapped spot. Fear of being branded by my fellow competitors as dishonourable. Guilty of doing something I shouldn't.

'Come on. What the hell are you doing in there? Get out of my stall,' demands the wheelchair pilot.

I frantically look for an escape route. Climbing over top of the stall would only land me in the wheelchair. As I peer under the stall to my right, I realise it's occupied. There's no escape route. I must open the door and face this irate, angry invalid.

My hands tremble; my throat dries; my stomach churns. I prematurely finish and flush. I slide the latch to open my stall door. As the door swings open, I assume the posture of an invalid, stumbling out of the stall and shaking violently. The occupant of the wheelchair looks stunned, mouth open, jaw dropped, and staring at me with an unbelievable scowl.

'I'm sorry to keep you waiting, sir,' I stutter.

I don't dare look back as I struggle to the sink to wash up. I hear the stall door close and I'm out o' there. Thank goodness. Due to the fact there was one empty handicap stall, I'm in time for the start of my match.

What's that Guy Doing in His Pyjamas?

It's my routine. Every morning at 4:00 a.m. I roll out of bed at our Mexican resort to head down to the beach. It's my duty to go and lay out the towels on our beach chairs to save our ocean-front palapa for the family. I'm not exactly sure how I got this daily task, but for over a decade now it's been my sole duty to make sure I get a choice spot. Six chairs, two tables and a clear ocean-front view across the sand to the rolling ocean waves in Nuevo Vallarta. Perhaps it's because I'm the only family member who can fall back into bed and go to sleep immediately until it's a sane time to get up.

My task done at the palapa in total darkness, I head back to our room. Since my routine is a short interruption in my nightly sleep, I never take the time to change. I scurry down in bare feet and pyjamas, ready to jump back into bed. My route takes me through a common building, across a well-travelled pathway, along a tree-covered path, and into my room. As I emerge from the common building, I hear voices of young people walking along the path.

Caught – barefoot with dishevelled hair – in my pyjamas. Too late to turn around! Too late to sprint out of sight! *What am I to do?* The fear of embarrassment flushes through my body. I don't want to face the ridicule of this young crowd. A thought flashes into my mind. *Get my arms up outstretched in front of me. Close my eyes. Stiffen up my body and keep walking.*

'Oh no, he's sleepwalking. We're not supposed to disturb him. We'd better be quiet and leave him alone.'

As I disappear down the tree-covered pathway back to my room, I can hear the chatter and giggling of that young group behind me. I believe it's the first time I've ever walked in my sleep. It's not as bad as they say it is.

Stepping into the Modelling World

I'm relaxing on the beach in Mexico when I get a call from an agent.

'Brad, I'm wondering if you would be available to be a model for an upcoming ad?'

'Absolutely, as long as the shoot is after I get back to Canada.'

'We'll spend a day taking photo shots for a promotion, and then we'd like to do a video with you. Would you be willing to take on the major talking role in our video?'

'You bet. Count me in.'

I'm excited about this opportunity to explore the world of modelling. I go to bed dreaming about exploring this new lifestyle. *Why did they choose me? What is the ad promoting? Who will I be modelling with?* I fall asleep with visions of posing with J Lo.

Today I spend time on the beach with my son Justin, a fitness and fashion model with a top international modelling agency. He promotes natural bodybuilding – no drugs. He fills me full of good advice.

'Dad, you've got to be confident – that confidence will come out in your photos. Be yourself. That's why they hired you. Don't try to copy anyone else. Exhibit your own style, a look no one else is capable of offering. Listen to and trust the photographers – they want to help you so they end up with the best photos. The more modelling you do, the more you will learn how to perform in front of a camera.'

***

It's the day of my first shoot – I'm filled with trepidation. As I enter the studio, I'm greeted by Roth and Ramberg photographer Tom.

'Hi. You must be Brad?'

'Yes, sir, that's me.'

'Great. Let me introduce you to Lucy, our fashion designer and Mary, our make-up artist. They'll get you ready for the shoot.'

'Can you let me know what the shoot is all about?'

'Sure. It's for a new seniors home. You're the perfect personality for their TV and newspaper promotion.'

So much for working with J Lo. They've just thrown my ego out the window. They say I'm perfect for a seniors home promo! I try to look brave and confident in my senior body as I get dressed.

What fun. It's a great shoot with the photographers helping me portray what they need. I do the video, which is screened for numerous weeks on TV stations. I start getting phone calls from friends asking when Bonnie and I are moving into the seniors home – one call is from an elderly lady who wants to know my room number! A lot of my friends and students say they hear my voice and then look at the screen to see my face. Who thought this professor would be a TV star?

Other promotional shoots follow. In one, I'm the father of a young farmer looking out over a field of grain ready to harvest. My facial expression portrays surprise with the abundant harvest, promoted by the advertising fertiliser company. My 'son' is a professional model flown in from Los Angeles for the shoot. I feel quite honoured to be sitting next to him in the pickup truck.

Another portrays me as a golfer. What? I'm no golfer! Bonnie and my father-in-law spend hours with me in our backyard, trying to perfect my stroke. After many retakes, I finally come up with an acceptable form, and they place my image on a trophy for advertising. Yep, you've got it - Bonnie finally has her 'trophy husband'.

I'm glad I didn't quit my day job. The modelling gigs are disbursed sporadically throughout my life, but certainly not enough to pay our mortgage. It's just another occasion of capturing an opportunity, embracing my fear and jumping in although I don't feel qualified. I love these uncomfortable occasions to learn and grow into who I am today.

REFLECTIONS. Quick Thinking

In life, there is no question each one of us will face dilemmas in which quick thinking could help develop the ability to think *through* fear, spelling the difference between success or failure, life or death. In my own life, I know there is no substitute for experience. Past experience is crucial. Those past experiences fill the toolbox with options to employ. With those tools, rapid and effective decisions are easier to make. In some cases, time is a huge factor. When effecting lake rescues, time is on my side since the victims are stationary. It's quite a different story with the victims being swept away by the current in a river rescue.

NASCAR drivers react in one of two ways as they witness a collision in front of them at 300 kilometres/hour. Some drivers hit the brakes and go into a four-wheel skid. Others instantaneously assess the situation and attempt to figure out a way to avoid it. Why are some drivers able to react in a calculated manner instead of panicking and hitting the brakes? It's a question I cannot answer. Are we born with the ability to think quickly, or is it an acquired skill?

Roth and Ramberg
Photography

Seniors Residence Model

Roth and Ramberg
Photography/Group 23
Golf Model

13. My Transformation Strategies

'In any given moment, we have two options: to step forward into growth or step back into safety.' **Abraham Maslow**

The tales I share with you are based on the reality I know, have personally experienced and share with total honesty – a life filled with curiosity, fear, terror and joy. I strive to explain the collision between ambition, fear and fulfilment as I live my full-contact life. I'm no expert, not a psychologist, not a self-help guru. I can only share my own experiences of dealing with fear and what they mean to me, with the desire of helping you mitigate your own fears. Gaining the courage to dance with fear is an enormous gift, transforming life beyond imagination. I honestly could die today without regrets. Perhaps some of my strategies will work for you. Take those and fine-tune them to your situation, your own risk tolerance, your personality and your mental toughness. Discard the rest.

Fear, like joy, is a natural biological human response when confronted with threatening events. It strikes like a lightning bolt and immediately changes our mindset. Joy fills our inner being with a sense of happiness. Fear is a warning system signifying danger. Fear knows not whether the situation is dangerous or something we can overcome. Each crisis offers two outlooks – danger or opportunity. Assessment is vital. In those fearful moments, we must remain calm, quiet our minds and build awareness so we can go into battle poised and in control. For me, welcoming fear signifies I must be fully focused on confronting the threat at hand.

```
    FEELINGS              THOUGHTS
         \                   /
          \                 /
           \               /
            \             /
             \           /
              \         /
               \       /
                \     /
                 \   /
                  \ /
                ACTION
```

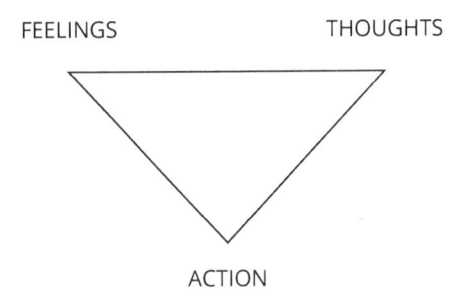

Many of my most transformative lessons have come amid my most terrifying moments – when I've learned to step into fear and take positive action. I agree with Dawn Huebner – danger encompasses thoughts, feelings and action. When thoughts and feelings control me, I freeze. I've got to step in understanding 'I'm afraid'. I must act, giving myself a chance to come out of this dilemma. Fear accompanies my action. Once I take that first step, my feelings disappear as I focus on my actions. As I looked up to sight the avalanche hurtling towards me, I discarded my skis and poles – my first step. After that, my full concentration was on surfing the slide. Fear vanished as I followed my life-saving practices. Similarly, when my canoe trailer came alongside, I took my first step of action – acceleration to get in front of the runaway trailer. After that, I was able to avert the disaster.

For each of us, exposure to fear varies greatly – from walking into a room of strangers to paddling amongst crocs. No matter what the danger, it can be either real or perceived. As an outdoor guide, I've learned that perceived fear is no less terrifying than real fear. We must dare to step through the threat we believe is present, whether it's real or not (remember the student with the trapped fish in her canoe).

Facing fear for anyone is serious and should not be taken lightly. Some of us are better at facing fear, but no matter what level of fear we witness, we must never laugh or make a joke of it. Acceptance and support are the right reactions.

Managing that fear is key. My own life has been full of terrifying moments, low and high on the risk scale. Facing those moments makes me stronger, more confident, more self-reliant, more resilient. They give me the courage to step into the uncertain, understanding that I'm building a toolbox full of strategies to manage fear. My mindset is crucial. Self-doubt can have no place in my thinking – fear loses power when I'm confident.

Every time I choose to step outside my comfort zone, I visualise myself coming out the other side successful. Visualisation is a skill I learned to use as an athlete and coach. It's the ability to visualise myself in triumphant action, executing perfectly in one of two forms. In one, I imagine myself as a video camera recording my movements. In the other, my mind is immersed in the actions – experiencing the sounds, smells and visuals as I 'feel' myself executing expertly. Although the second skill is more difficult, I find it more beneficial.

Understanding the magnitude of the danger, my own skill set and mental toughness all contribute to my decision of whether to become engaged or not. Is this an opportunity to learn more about myself and the world, or are the consequences tragic, scarring me for the rest of my life? I've come to realise the more experience I have, the more I should follow my gut feelings. Past experiences, both cognitive (those rehearsed in my mind) and experiential, are crucial in making sound decisions.

As I face fear, I've developed a game plan – principles to help decide if I should step into the challenge.

Where is this fear originating?

Why do I feel fear – what is the root cause of my fear?

What am I afraid of – is it real or perceived?

What skills do I possess to meet this challenge – how will I deal with this fear?

Just as I coach, developing a game plan is crucial for success. My volleyball game plan includes: scouting the strengths of my opponent, then matching my strengths to mitigate those game-changing factors; identifying opponent weaknesses that can be exploited as I plan to take advantage of that Achilles heel. Remember, 'a weakness is not a weakness until it's been exploited'. Similarly, preparation, with a solid game plan, is crucial before jumping into a fearful endeavour.

In my life, transformative events have occurred in different ways. *Forced actions*, designed by another caring person who believes the experience will be beneficial. *Endorsed actions*, engaged in with my own desire to participate, believing I will grow through the experience. *Unplanned action*, when an activity turns sideways, facing failure unless I'm able to overcome the danger. As James Lane Allen claims, 'Adversity does not build character, it reveals it.'

Forced Actions

Most of my 'forced actions' occurred early in life – a lot of them by my parents. My parents designed scenarios which forced me to battle through my fear. I had no choice but to accept fearful actions when told to shake hands, step into the shower, portage a canoe and go out for a game-winning faceoff. My actions changed the way I thought. Fear reigned when confronted by the task. However, I dared to act since I believed those in charge cared for me, that I could accomplish the desired outcome, and knowing I was supported.

> *Get ready to accept that which you allow.*
> *Fear leads to sub-standard behaviour.*

Accepted Actions

Endorsed actions were those adventures I willingly entered, believing I could create a more enriching life for myself – going beyond the way I live – stepping out of my comfort zone. I fully accepted the consequences of these actions as I voluntarily plunged in. I believe these experiences help unearth a profound sense of self – swimming with a humpback whale, kayaking the Sea of Cortez, feeding crocs with a Juju priest and accepting international pro coaching positions. It's essential I never stop growing as I explore my human potential with trusting relationships and exhaustive research, believing I'm competent. These stretching events illuminate the purpose my Creator has placed me on earth to fulfil.

Perhaps my insatiable *curiosity* is the main impetus in overcoming fear. My desire to fully experience new adventures, gain insights into my core self, unearth how far I'm able to stretch before breaking. Dweck terms it a 'growth mindset'. The emotional rewards of my endeavours are without match. They are surreal – beyond joy. Once experienced, the courage to tackle the next challenge becomes easier. On rare occasions, they may change the course of my life. *'At this fork in the road of my life, I choose to return to Canada, preparing to start Canada's first Outward Bound School.'*

> **Does this activity align with my passions, placing another piece of the puzzle into realising my life game plan?**

Some decisions are easier than others – the consequences are more black and white – *taking action is a no-brainer*. Refusing to act, as fearful as it may be, is simply wrong. When diagnosed with cancer, I fully understood the possible side effects – incontinence, impotence, perhaps death on the operating table. The other option, dying *because* of cancer was not on my bucket list. I had to take immediate action. *'Please book the surgery.'* Fortunately, the skillful surgeon saved me the humiliating side effects.

> **Fear the consequences of no action, they could be catastrophic.**

The fear of ending my father's life was a heart-rending, grim, but *merciful decision*. Emotions run amuck any time we make the decision to take a loved one off life support. My decision was certainly made more easily since I'd spent such a rewarding week with Dad, and the very clear, absolute desire he shared with my sister and me. Even though my decision, *'Take Dad off life support'*, could barely come out of my mouth, I categorically had to make it.

> **Sometimes the hardest decision is the most fearful, but the correct decision.**

As I face fear, my brain fills with questions. What might happen? Should I go ahead? What's the upside? I've found many of my strategies require *intentional training* – skill sets that must be developed over time using my thoughts, visualisation, faith and actions. I strive to listen to positive outcomes rather than negative – 'I am enough.'

> *Intentional training will fill my toolbox with options to overcome fear.*

Sometimes, *one fear may override another.* The fear of jumping into an activity is eclipsed by the fear of losing that once-in-a-lifetime experience. While diving in the Red Sea, the ranger stated, *'I'm sure you will encounter sharks while diving.'* Don't you think I feared meeting a shark, unprotected? You bet. But I was more afraid of never getting this chance again, something I'd dreamt about doing. The ranger gave us a strategy for surviving. Count me in.

After suffering from sciatica for five years, I decided to undergo major spine surgery. *'I refuse to live the rest of my life like this.'* Yes, the outcome could result in paralysis, but I focused on the positive probability of regaining full use of my legs.

At times, trepidation will rob us of amazing experiences. When Bonnie and I were in Venice, I wanted to treat her to a romantic gondola evening cruise. I couldn't justify the expenditure, and we missed out. My bad – that opportunity may never present itself again.

> *The thought of missing this opportunity may outweigh my feelings of fear.*

Surrounding myself with positive, *like-minded individuals* is critical. When sharing ideas, these supportive friends are crucial. Waiting for our turn to capture a wild boar in Italy, Bonnie offered advice rather than negative declarations, *'You'd better get warmed up.'* Her support fuelled my confidence.

> *Sometimes supportive, like-minded friends are the only ones who understand and encourage brave decisions.*

Sea kayaking a dangerous circumnavigation in the Seas of Cortez without a guide was a challenge not to be taken lightly. I carefully *assessed* my own relevant skill set, and those of my son and daughter-in-law. Each of us is highly experienced when it comes to outdoor travel and survival. My conclusion: *'We are capable of facing these challenges.'*

> **Astute assessment can override fear. 'I am enough.'**

There is no question my *faith* helps me be brave enough to dare to tackle fearful decisions. I firmly believe a Higher Power plays a role in the outcomes of my actions. As I've said, *'My miraculous result can only be described as Divine interaction.'* Can I prove this? No, but I can point to occasions when I have no other viable explanation. My recovery from so many deadly medical events defies reason. On the death of my two sons, my grieving was immense. However, I was bathed in a peace I cannot understand. Escaping epic adventures that could have ended my life with no explanation leaves me baffled.

> **I have no explanation for the miraculous way I survived, except that a Higher Power is looking after me.**
>
> **'Even when I walk in a dark valley, I fear no evil because you *are* with me.' (Ps 23:4 LEB)**

It's a state I seek in competition. Good results gain more success. Those occasions when I come through the storm successfully build *confidence* – that I'm able to do it – that *I am enough*. Facing fear head-on, I reflect on ways I've been able to handle fears in the past. As a river rescue expert, my expertise was built with sequential, progressive, experiential mishaps building a repertoire of skills. *'Brad Kilb's experience in the field of canoe instruction, white-water rescue, and self-rescue is without peer in the world today.'* I discard my doubt as I rely on past successes to guide my actions, as fearful as they may be.

> **Success, when dancing with fear, builds confidence.**

It's important to understand why we feel fear. Often it could be the fear of *shame* which could arise from losing, not being good enough, being embarrassed. Yes, I've felt that shame and embarrassment as I competed – both as an athlete and coach. After losing badly to a Hawaiian team, I faced the athletic director. *'Your team is an embarrassment.'* Quitting is out of the question. There is no shame in learning from our setbacks. I know I'm capable of correcting those shortcomings.

> *Fear challenges me to learn of my weaknesses, and I vow to correct them.*

I've witnessed perceived fear in the lives of others – the student in the swamped canoe with the entrapped fish. I often could not *understand that fear*. I decided to embark on an expedition to intentionally put myself in terrifying situations. Bonnie and I rented a twelve-metre yacht and sailed up the British Columbia coast with our two young boys. Although there were many frightening encounters, the most petrifying was getting caught in a raging storm with heavy seas. Forcing myself to focus on the fundamentals helped me understand how to assist others overcome their fear.

> *Be bold enough to experience and feel the fear of others.*

Pulling our canoes onto the shores of the Ram River following our epic expedition with the graduating outdoor pursuit class was indeed a relief – relief that we'd dodged any calamities. Bill March and I decided we'd never lead such a *dangerous* trip again – it was simply too life-threatening.

> *Sometimes fear can assist in grasping the severity of the undertaking - say 'no'.*

As I decide to tackle a new adventure, I must occasionally *ignore sceptics*. They may deprive me of an opportunity to grow. I know Darrell was concerned for my safety as he warned against entering the water with a humpback whale. *'Brad, don't do it.'* In those brief moments, I had to decide who was more knowledgeable, the expert whale researcher or Darrell.

> **The trusting relationship of a knowledgeable partner will help make wise decisions.**

Posing for an advertising agency can be nerve-racking. Portraying an acceptable image can fill our thoughts with, *'Am I really good enough for this shoot?'* As my son pointed out, *'Dad, you've been asked because they see who you are. Just be yourself.'* This truth – understanding I'd been recruited for who I am – helped build my confidence in front of the camera. Sometimes, others can see more value in us than we can see.

> *'The privilege of a lifetime is being who you are.'* Joseph Campbell

Walking into my principal's office could spell my termination. I didn't have to take the scary action to share my drug-related revelations about his daughter. I made my decision based on *moral principles*. My integrity hung in the balance. There was only one decision to make, as frightening as it was.

Did I always make the most ethical decision? No, I did not. I can think of times when I made the easy decision, the non-threatening decision, the wrong decision. I can't say I'm proud of all the decisions I've made. When coaching in the World Volleyball Championships in Brazil, I stayed behind to observe the Argentine team in their pre-match 'closed' practice. Yes, it was unethical. I cheated. I should not have done it.

> **Fully understand your personal values and never abandon them, even if you fear the consequences.**

'Brad, would you please coach our team this season?' Fear filled my cranium when asked to coach one of the top teams in Canada. I simply didn't possess the knowledge or experience to coach such a high-level team. It was only after they convinced me that I had the skills they required, and they would support me in areas of weakness, that I agreed to coach. What an amazing growth experience.

Unfortunately, I feel our society looks upon *vulnerability* as a weakness. In my experience as a leader, I've found I gain tremendous respect when displaying vulnerability. Expressing an apology, not knowing all the answers, sharing that I've made a mistake – all make me human. It's amazing how the more vulnerable I become, the easier it becomes. No shame. No guilt. No fake words. Less fear.

> *Vulnerability is an essential trait of strong leadership – it's our greatest measure of courage.*

Comedy can sometimes be the best cure for fear. Easier said than done! Being deluged with panic can cause us to freeze – verbally and physically. When caught in my PJs, my ability to insert fun is what saved me. Striking the pose of a sleepwalker allowed me to escape that moment. I believe quick thinking in stressful scenarios is a skill that can be developed through practice. Often, I'll think of a comical response I could have made – too late. Tuck that away in your brain. You may be able to use it in the future.

> *Comedy is one way to release tension and overcome fear.*

I remember a day paddling a very technical piece of white water. *'I'll bet nobody can paddle that drop.'* The taunt was too great. My *ego* got in the way, and I paid the price – floating downstream with my shattered canoe. Listening to my own assessment should override dares and ego.

> *Don't let ego get in the way of wise decisions.*

Unplanned Actions

Unplanned actions occur when an activity turns sideways, with fear of the looming disaster – surfing when caught in an avalanche, fighting an armed thief, leading river expeditions. These unexpected, fearful events teach me to remain calm, focusing on the moment as I concentrate on fundamental skills learned through previous training and adventures.

There's no shortcut when it comes to *preparation*. Developing skill sets required to be successful in stressful scenarios demands practice. Filling the toolbox with those necessary skills is imperative.

> *Confidence built through preparation will help you take on fearful actions.*

My advice to athletes in times of intense battle is always the same. 'Try to relax and *focus on the fundamentals.*' As I planned my escape route from the menacing croc, I played those words over and over in my panicked mind. As I convinced myself to think of those basic strokes, my fear diminished.

> *Fear will diminish as we focus on the fundamentals.*

In a battle with *nature*, I will always lose. Instead, I must ignore the urge to fight, but rather find a way to work in unison with a power much greater than my own. When caught in an avalanche or river hydraulic, I must understand what forces I can use to my advantage. Understanding how to utilise nature's forces may help us escape disaster.

> *Sometimes, nature may release her fury upon us. Harness rather than fear those forces.*

As I became conscious in the intensive care unit, I heard the words, *'Sir, can I pronounce the Last Rites for you?'* What a numbing, terrifying statement! I'm not ready to die. Are you kidding? It's not my time. Obviously, I have more faith than you. *'No, sir. Do not read me my Last Rites. I've got lots of living yet to do.'*

I refuse to give up. I'll do everything possible to battle this. My Creator will keep my breathing. He has work for me to do.

> *Although it may appear as if fear is winning, never, ever give up.*

Occasionally I've come across *someone in serious trouble*. The instantaneous decision of whether to act and help is full of fear. Scrambling over a hilltop to witness two men trapped in a downed plane resulted in an immediate response. *'We've got to try to get that pilot out of the inferno!'* Perhaps it's the adrenaline that assists us in those situations?

Watching my business partner roll on the floor with a grenade-wielding drunk, I had to make up my mind – fight or flight. *'Brad. Help me. He's got a hand grenade!'* My care for Darrell superseded my fear of being blown to shreds.

Coming across Dad, unconscious on the kitchen floor, demanded action. Do I follow his instructions – *'I don't want any heroic interventions?'* My gut told me what to do, Call 911.

> *Occasionally I must rely on my skill set to act without fear.*
> *Don't hesitate when someone's in trouble.*

There've been times when I just *didn't know what I should do* – times when I've never confronted the situation before. Looking through the open shelves in our liquor store to witness a thief in action filled my brain with fear. I had no game plan. I only knew I must do something. Leaving the safety of the shelving and approaching the shoplifter was the best action I could start with. From there, my natural instincts took over – extending my hand for a handshake. *'Thank you for coming to our store.'*

Hearing the commotion in the schoolyard while on supervision duty was another situation leaving me with no answer. I was scared. What to do? Blindly starting to proceed without any defined action was the best thing to do. Doing nothing would not resolve anything.

> *Once I take that first frightening step, other actions will often fall into place.*

In my various *leadership roles*, stepping up even when filled with fear is vital. Our followers look up to us to act responsibly, to set the example, to take steps to mitigate and respond to misadventures. I believe we must *not* jump into an adventure without knowledge, skill, extensive prep and mental toughness. These attributes can be trained and help, not guarantee, success.

When caught in a misfortune, my thoughts revolve around not only myself, but also those I'm leading – as when a hurricane strikes when teaching canoeing, when paddlers are in dire need of rescue, when my family looks to me to bring our yacht safely to port. While my first response may be to escape the disaster, I must expand my thinking as I struggle to ensure everyone is safe.

> ***Leadership demands effective action,* even when filled with fear.**

The challenge of making the right decisions in stressful situations can sometimes depend on not what we know, but *how we use what we know*. As the coach of our volleyball teams competing in the finals of a national championship, I had spent time garnering information on our opponent – my scouting report. However, that info was of no value until I decided how to use it to exploit my opponent's weaknesses – my victorious game plan. I maintain, 'A weakness is not a weakness until it's been exploited.'

> ***Success depends upon how we use what we know.***

14. Don't Let Fear Limit Our Potential

'Fear defeats more people than any other one thing in the world.'
RALPH WALDO EMERSON

My life took a dramatic change when I attended the Nigerian Outward Bound School. It was indeed a transformative event. The realisation I could choose an occupation in which I'm able to live out my passion every day was a revelation that literally changed my life. The fulfilling life I've lived for decades is a result of that decision – to *pursue a lifestyle* in which I'm fully able to utilise my God-given gifts. I've found my purpose here on earth. It's not to accumulate financial wealth or lavish possessions, but rather to facilitate the growth of leaders through my mentorship.

I tell my students, 'I'm not a financially wealthy individual, but *I am a multimillionaire when it comes to my lifestyle.*' I wouldn't change it. It's a lifestyle that makes me whole. I love my job. It's why I awake every morning and leave home excited about my career. I can say I'm unconditionally comfortable in my own skin, so grateful for the God-given gifts I've been endowed with. I believe they enable me to welcome fear as I live a truly authentic, vulnerable and humble life – qualities I revere deeply.

Now, in my senior years, I don't want to become famous, rich or win more awards. I want to enrich relationships – to impact the folks I engage with in a positive and facilitating manner. My challenges are changing – more age-appropriate but still with a teachable spirit. My values, philosophy and behaviours are the result of decades of evolution and transformation based on real-life experiences. I don't want to create boundaries constructed by fear which limit my potential. I endeavour to continue living an exciting, fulfilling life until they truck me off to the crematorium. For me, that beats living a schedule filled with Netflix, neighbourhood gossip or the country club.

How can I say that I have a love affair with fear? I've discovered that fear *is* my friend – it has peeled back so many undiscovered layers about myself and our world. My evolution has been the result of *welcoming fear into my life* – daring to accept that fear is perhaps the greatest transformative partner resulting in personal growth. I pray I will continue to have the courage to face fear head-on until the day I pass, knowing that I've lived life the way my Creator designed.

Ever So Grateful

I start by thanking you, my family, friends, students and athletes who encouraged me to record my outlandish stories. 'Brad, you've got to write a book about your adventures.' I'm so appreciative to all of you who have made it possible to share my story. Each one of you has exhibited your genius in different ways. Your diversity of gifts has resulted in taking my storytelling and vividly converting my yarns into this memoir.

How could I start thanking my team without first acknowledging you, Bonnie, my wife. As a creative professional artist, you've accompanied me on my journey. Your encouragement, caring attitude and insights during our dark days have enabled me to produce my story. You understand me in ways no other has, as you inspire me to utilise my God-given talents to follow my dreams, face my fears and develop skill sets. Your willingness to jump into the mayhem with me has made my life complete. God could not have given this brave adventurer a more empowering partner.

To my six kids, each one of you is so different yet so willing to uplift me through my successes and failures. Your continual challenges help me to become not flawless as you well know, but the best I can be. Thanks Jamey, Bryn, Jodi, Brad Jr, Brett and Justin – I love you so much. The world tells parents we should be the example we hope our children will grow into. For me, I strive to demonstrate the example of living life to the fullest with integrity. A special heartfelt appreciation to you, Brett and Brad Jr, who passed away far too early, teaching me the importance of having fun as I strive to live life to the best of my ability, regardless of what it is I choose to do.

To you select few who have given of your time to read my manuscript and provide valued feedback – on the beach under our palapa in Mexico, on our patio around our firepit during COVID-19, over coffee in a favourite café – you have assisted me to write the stories of transformative changes because you know me. A genuine 'Thanks' to Tal Guppy, Colin Kubinec, Loral Kilb, Justin Kilb, Bryn and Christine Kilb.

To my students, athletes and coaching teams – you've taught me so many life lessons. You call me 'coach', but I often felt it was you mentoring me.

To my patient and expert editing team at Friesen Press - Laura Matheson, Teresita Hernandez-Quesada, Oriana Varas, and my Publishing Specialist – you've held the rudder as we've charted the course. You have been so professional to work with. Your awareness of what readers look for has been invaluable.

Writing my memoir would not be possible without you, Mom and Dad. The full-contact life that you lived created within me a direction I've never regretted.

Finally, to God who, without His gifts, guidance and loving Presence, I would be unable to live my life as He has designed it – a life of dancing with fear.

About the Author

Cowboy Brad
Artist: Jody Skinner

Brad's researched fear as an educator for fifty-one years, traveling to sixty-seven countries, residing in six. He's a University of Calgary *Teaching Excellence Hall of Fame* inductee; coauthor of a Canadian *Best-Seller*; *Gold Medal* recipient the at the Banff International Film Festival; served as a *Technical Official* at the London Olympics; awarded Canada's University Volleyball *Coach of the Year*; coached the *Canadian National Volleyball Team* and professionally in Italy, Switzerland and Japan and has won more *National Championship rings* than he has fingers.

Brad lives in Calgary, Canada with his wife Bonnie MacRae-Kilb (artist and former National Team Volleyball Player). He loves outdoor activities, mentoring, and West Coast carving.

bradkilb.com   kilb@ucalgary.ca

CPSIA information can be obtained
at www.ICGtesting.com
Printed in the USA
BVHW020251220622
640342BV00018B/170